AF797942

LIFE AND LIVELIHOOD OF THE MARGINALISED

MJP PUBLISHERS

LIFE AND LIVELIHOOD OF THE MARGINALISED

G. Palanithurai

Professor, (Retd.,)
Rajiv Gandhi Chair for Panchayati Raj Studies
Department of Political Science and Development Administration,
Gandhigram Rural Institute - Gandhigram,
Tamil Nadu.

MJP PUBLISHERS

Chennai **Trichy** **Tirunelveli** **New Delhi**

MJP PUBLISHERS

ISBN 978-93-88694-77-3

MJP Publishers
No. 44, Nallathambi Street,
Triplicane, Chennai 600 005

Printed and bound in India

MJP 769

Publisher : **C. Janarthanan**

Project Editor : **C. Ambica**

lematic. Hence, I have conducted a study in the implementation process of Manual Scavengers Prohibition and Rehabilitation activities. In this small monograph, three issues have been analysed at micro level. They are Manual Scavengers, Domestic Workers and Mayana Vettiyans. Their life and livelihood issues have been analysed at micro level. The epitom of reports have been published in the journals. They have been sent to the appropriate officials for further action. Though all the three are micro analysis the results and explanations reflect the general reality. If the policy makers are interested, they can act upon the findings of these studies. Fourth chapter deals with the rural realities from the perception of the youth organizers in the rural areas. Since they work with youth in the villages and they know the pulse of the rural. Hence, a study has been conducted. All the four chapters are published in journals as articles and based on the studies, popular articles have also been written and published in reputed newspapers.

To

Divine Mother

Preface

Ever since the Arjun Sengupta committee on unorganized sector subm ted its report and discussed in parliament, I have been obsessed with t research activities carried out in the higher learning institutions in In which are irrelevant to realities of the society. Most of the research act ities are career building activities not pertinent to solve any of the pre ing social problems. In the same way, the world inequality report, wo gender gap report, world hunger report, world development report, wo human development report, world livelihood report, and world hum rights report have increased my obsession. Further, my conviction the theory of state propounded by M.K. Gandhi has been increased. one stage, the state itself declared that it could not fulfill the wants of t people and thereby market has taken responsibility. But now market h also declared its inability to address the issues of the people. Since I w influenced by the scholars belonged to post behavioural revolution, fro the very beginning of my research career I started working on socially r evant issues. I won't stop with production of reports. I will take it to t appropriate bodies for making appropriate decisions and policies on t critical issues of development. I won't stop with that and move furth I used to involve mainstream media to disseminate the findings of t research projects for the popular consumption. This is my framework action in research. In this regard, I have taken up a few studies which a highly relevant to the context. Government of India has been conte plating a legislation to protect the job and rights of the Individuals w are working as domestic workers. Now an Act has been passed. In t same way, Manual Scavengers Prohibition and Rehabilitation Act ha been passed in Indian Parliament. Implementing the Acts becomes pro

Acknowledgement

When research scholars are selected for Ph.D programme I used to encourage them to take up some critical socially relevant issues for investigation. I normally treat Ph.D scholars as co-learners. In our department, it is a practice that research scholars have to work on socially relevant projects as it is primarily a Gandhian institute. Once research scholars are selected for admission, I used to interact with them with an objective or orienting them to take up critical development issues for investigation. In this process, two of the researchers in our Department helped me in collecting data for preparing papers for presentation in different conferences. They are, Mr. G. George and Mrs. Aarthi and both are Ph.D scholars and they have been entrusted with the responsibility of collecting data from the field on all the three issues which I have taken up for investigation. G. George is doing research on Manual Scavengers Prohibition and Rehabilitation Act and Aarthi is doing her research on Mayana Vettiyans. I have asked both to collect data from the field. Focus Group Discussions I have conducted in the field. Some of the narratologies have been done by me. Both G. George and Aarthi have to be complemented for their help rendered to me. In this process they also got clarity on many of the aspects of these issues since they are doing their Ph.D. on these issues.

Next I have to thank the editors of the journals for permitting me to publish the same. My colleagues are always behind me in the Department for all my academic activities and they deserve my appreciation and compliments. In my 29 years activities in the Department I found always harmony among my colleagues. They all value my work. They enabled me to move in the direction as I want. It is a Political Science Department where there is no politics. They are all highly matured and hard working teachers. Hence, I register my sincere thanks to all my colleagues. Last not the least my family unit, my wife and children. They know the value of my work and hence they value me by allowing me to spend the whole time for my academic research and outreach activities. Hence, I register my sincere thanks to my wife Mala, Daughter Senthil Vadivu, Son-in-law Senthil Murugan, Grand Son Pranav and my son Aurobind. My research associate D. Sivakumar is always generous enough to cooperate with me to work longer time in the office to carry out my works and he deserves my appreciation.

G. PALANITHURAI
Gandhigram

All the chapters in this monograph give the readers a glimpse of the issues associated with a few segments of the society. These are the tips of the iceberg and large number of such critical issues has to be studied.

NOTES AND REFERENCES

1. Arend Lijphart (1991), **The Puzzle of Indian Democracy: A Reinterpretation**. RGICS Paper No. 18. New Delhi: Rajiv Gandhi Foundation.
2. Madhu Purnima Kishwar (2005), **Deepening of Democracy: Challenges of Governance and Globalization in India.** New Delhi: Oxford University Press, Pp 1-61.
3. Amrita Basu and Atul Kholi (eds) (1998), Community Conflicts and the State in India. New Delhi: Oxford University Press.
4. Gvenville Austin (1994), **The Indian Constitutions' Seamless Web**. RGICS Paper No.15. New Delhi: Rajiv Gandhi Foundation.
5. Sadig Ahmed (2007), **Job Creation and Poverty Reduction in India: Towards Rapid Sustained Economic Growth**. New Delhi: Sage Publications.

Contents

Chapter I

New Narratology of Women Domestic Workers in Tamil Nadu: A Micro Analysis

INTRODUCTION

In Indian society much of women's labour is informalised, unvalued and unrecognized. It is being done unconsciously and intentionally as it warrants establishing control over women by men. Women and girls involved in domestic work remain unorganized and unnoticed as poor are involved in such activities in order to meet the challenges of poverty. Since they are under the clutches of the middle class, upper middle class and the rich, the policy community paid poor attention on the issues faced by domestic women workers. As a result, studies are scarce, lack of data base and poor attention of the media on the issues of the domestic workers. In general domestic work never gets primacy among the policy makers as house is not considered as work place. But finally the conditions of the domestic workers have been captured through a study by the Government of India (GOI, 2011). Yet there are attempts to organize them as groups or unions and they are also not large in numbers (Bhagyashree and Hilaria, 2016:1-8)). Even though the activities are lesser intensive, the state responded to the plight of the domestic workers through a new mechanism. It is nothing but inclusion of the domestic work in the list of scheduled employment under Minimum Wages Act of 1948 that too in a few states (ILO, 2012). Even states which are progressive and known for gender activities have not sufficiently paid attention on the issues of the women domestic workers (Neetha, 2008: 26-28). Individual attempts had been taken to ameliorate their conditions through legislation in vain (Sonia, 2013:69-76). The media which are active both print and electronic have not covered the stories of the women domestic workers as they have gained neither orientation nor sufficiently conscious of the plight of the women domestic workers. The issues and problems of the women domestic workers vary from state to state and district to district within the state.

A "domestic worker" is a person who works in or for a household and where there is an employment relationship. These workers can be local, migrants, working in one or many homes on a daily basis or only for certain days in a week., but regularly for their livelihood. It is the definition given by ILO convention 189. The convention recognizes the rights as other workers such as right to collective bargain, security, minimum wage. A domestic worker performs variety of household services for an individual or a family, by providing care for children and elderly dependent to housekeeping, and addition to this they use to clean the house and the households, washing, cooking, other than this laundry and ironing, shopping for food and so on.

PROBLEMS OF THE DOMESTIC WORKERS IN GENERAL

Majority of domestic workers are women, who all are working in the household. In India, there are about 16.8 lakh female domestic workers in the country and of them 6.2 lakh are males (Bhatt, 1989). Each woman has different pathetic stories, which may explain the reason behind the field they choose. In India, 45% of women are illiterate, so definitely, they have to work in unorganized sector for their livelihood without the awareness of their rights and entitlements.

LEGAL PROTECTIONS

The United Kingdom's Master and Servant Act 1823 was the first of its kind and influenced the creation of domestic service laws in other nations, although legislation favoured employers more than of servants. However, before the passing of such Acts servants and workers in general, had no protection in law. The only real advantage that domestic service providers was got meals, accommodation, and sometimes clothes in addition to a modest wage. Service was normally an apprentice system with a provision for advancement through the ranks.

At its 301st Session (March 2008), the ILO Governing Body agreed to place an item on decent work for domestic workers on the agenda of the 99th Session of the International Labour Conference (2010) with a view to the setting of labour standards. The problems faced by domes-tic workers varied considerably throughout history and in the contem-porary world. In the course of twentieth century movements for labour rights, women's rights and immigrant rights, the conditions faced by do-mestic workers and the problems specific to their class of employment have come to the fore.

In 2011, the International Labour Organization adopted the convention concerning decent work for domestic workers. In July 2011, at the annual International Labour Conference, held by the ILO, conference delegates adopted the resolution on domestic workers by a vote of 396 to 16, with 63 abstentions. The convention recognized domestic workers as workers with the same rights as other workers. On 26 April 2012, Uruguay was the first country to ratify the convention.

In India, National Domestic workers Movement was started and mobilized the domestic workers and it gained visibility only after regional groups formed in different places such as Mumbai, Patna, Chennai, Varanasi, Andhra Pradesh, Kerala, Karnataka, Madhya Pradesh and Jharkhand. Though they formed groups and started mobilizing the domestic workers, formally they put up demands to state governments to enact legislations to protect them only after 1990. Because, they did not get adequate data for policy advocacy. Despite all such small mobilization, it is a pity that "unorganized sector workers social security bill 2005" did not find any mention of the domestic workers issue. Karnataka is the first state to fix minimum living wage for domestic workers in 2004. Subsequently many states took efforts to enact legislation (Neetha, 2013:77-84). Yet, conditions are appalling and deplorable (Hamid, 2016).

The issues are much complicated as they are in the clutches of the female family head. Even though , there are legal frameworks to protect them and safeguard the interest of women, their intricate working conditions made them on the job as half slaves in a free society. In the above backdrop an attempt is made to study the life and livelihood conditions of the domestic female workers in rural setting in Tamil Nadu. It is part of a larger study. The study is undertaken with a following objectives: a, under what family background the women are driven towards such domestic works in the families; b, what are all the activities carried out by them regularly; c, what are the economic and other benefits they gain for the work they do in the families; d, what impact the domestic works create in the families in terms of transformation; e, what are the problems faced by the domestic women workers in the families they work; and f, what are the enabling conditions needed to improve the working conditions of the women domestic workers.

METHODOLOGY

In order to investigate the problem case study method is being administered. Within the case study method, narratology, a nuanced tool is being adopted to get full details of their family stories, working conditions and the problems they have faced in their families and the families they work as servants, and the intensity of issues arising out of the men folk in the families where they work as domestic workers. By adopting the nuances of narratology, data pertinent to their personal problems are being collected. To complete a case study, the scholar has to devote a few days. The scholar has to create a setting for data collection by working hours and days together with the respondents. Data have to be collected from the full personal story of the domestic workers. With a promise given to the respondent to maintain the secrecy data are being collected from them. One case study is almost one full story about one individual.

Researchers have to create a rapport with the domestic workers and build a confidence in them that the data shared with the researchers will not create any problem either in their families or in the families where they work. Their integrity and name should not be tarnished with the data drawn from them. In such a way, rapport was created with the respondents to collect data from them. Data have been collected from the respondents in different spells in a week time. Even to collect data from one respondent, the investigator has to visit her many times in a week. The researchers involved in data collection have been trained and oriented by the author of the article. Data collection process by using narratology itself becomes a subject for deeper analysis. From the data collected, they are being analysed quantitatively. From the narration of the stories certain main points have been taken and they have been presented in quantitative table form for easy understanding. They are being supplemented with full narration of few cases. All the tables are in appendix.

DATA ANALYSIS

From the tables first ten in the appendix, one can easily infer that however poor they are, they got basic facilities to lead a decent human life. Further, through the development schemes and institutions meant for delivering goods and services to the people, they got their entitlements and services properly without much difficulty. Since Tamil Nadu is a pro-poor state and it has got many pro-poor schemes, the poorest can also lead a decent human life if their life is organised properly and avail the schemes and programmes periodically and properly. Tamil Nadu is known for its infrastructure facilities and hence they are being used profitably by the poor for their livelihood. In Tamil Nadu, poor are highly empowered.

From the tables 11 to 19, one can infer the rising revolution of aspiration to educate their children. They strongly feel education is the only instrument through which one can uplift the social and economic status of the family. Everyone believes that education is a powerful instrument to achieve transformation in the family and society and more particularly in the financial conditions of the family. It is considered as transformative engine. Their handling of money, use of the banking system, developing savings culture, investing the saved money in education of their children have lot of stories to tell about their vision, hope and aspiration. Not even a single maid servant wants her children to work in the sector as they work. They never considered it as a profession to remain in the work as other works. They opted to do this work as there is no alternative to them. They are trying hard to educate their children. They are courageous to spend huge hard earned money in their children's education. Apart from their savings they borrow money and spend it for their children's education. It is to be noted here that majority of the respondents never indulged in borrowing money for interest. Equally, it is interesting that the respondents follow small family norm.

The tables from 20 to 23 give the picture about the nature of work the respondents do in the houses, how they do the work, timing of their works and how much earn through the works. The nature and type of works and the amount they earn from their works have been projected. It all vary from individual to individual. There is no uniform wage for the nature and type of work in every family. In the same way, there is no uniform wage for the quantum of work they do. It all depends on the bargain between the worker and the household and in the same way it depends on the relationship between the worker and the household. No way the act meant for the workers has helped them in fixing their wage and the conditions.

The tables from 24 to 31 explain the nature of relationship between the workers and the households. As long as, both maintain a cordial relationship, the workers are not treated badly and shabbily. Majority informed that the household had concern for their domestic workers. It means they value the work of the workers. They know the importance of their work also. Yet there are abuses in the houses and it depends on the nature of relationship develop between the workers and the households. They are pushed to this work by poverty and thereby their aspiration to earn more money for their future and for their children. It is a pity that they are all in SHG and yet they are not aware of their entitlements. Majority expressed that they are so tired of their work. They work because of

the conditions compelled them. It is unequivocally clear that they are not happy with this job. They are in this job out of compulsion. Their poverty pushed them to this work and their aspiration made them to continue in this work. But their health conditions make them to be free from this job. They are unable to get out from this job as they need money.

The tables from 32 to 36 give glimpse of the family conditions with which they perform their work. They do not have their own association. They found uneasy relationship in their families despite their hard work and earnings. They face lot of problems in their families either because of their husbands or because of their in laws. Mostly their husbands are irresponsible and they are the main source of trouble. On most of the occasions their hard earned money is being wasted only by their husbands. Being women they are moving with all age groups in the families they work. Sometimes they get instruction from the male members and act accordingly. Their movements with men in the families where they work is a source of conflict in families of the domestic workers. Since they are not able to perform their family work they have to face the wrath of many in their families. Everyone wants money of the domestic workers but at the same time they have to work in their families also as domestic worker. This is the pitiable part of the patriarchy in our society. Most of the problems in their families are raised by their husbands. The domestic workers bargain with the house owners to fix their wages and the service conditions. It is not done in the backdrop of the laws or policies. But through their strategies and they settle their problems with house owners very smoothly.

While going through the tables from 37 to 39, one would find that the living conditions of the domestic workers are not so bad as government has provided needed facilities to lead a decent human life. But the working conditions are not conducive to lead a dignified life as we find in other places (Kundu, 2007:853-866). Apart from the above salary, working conditions, nature of work, intensity of work, health conditions, gender related problems vary from place to place house to house as we see in other states (Coelho and Chandira, 2013:39-46). At one point of time, there was no adequate data but now there are macro studies (GOI, 1988) and micro studies Kiran, 2013:63-68) to portray the conditions of the domestic workers. Yet, there is no comprehensive law for domestic workers in India (Kamala, 2013:85-89). When it is women domestic workers, the conditions are appalling and vary from city to city and province to province.

NARRATOLOGY

Case No: 1

Mahalakshmi (name changed) is 72 years old lady still engaging herself as a maid servant for her survival. She was born in Thiruthani , and she belongs to BC Yadav community. Her husband worked as the fitter in Thiruthani. She gave birth to eleven children. Her husband migrated from Thiruthani to Dindigul for want of job. At the age of 44 Mahalakshmi lost her husband. At the time of her husband's death only one son was with her and all others have got married and settled in different places. The son who was with her became a drunkard. All her children are illiterate and they are all going for daily wage in different places. Absolutely, they are not in a position to take care of their mother. Hence, her sons are not in a position to help her for her livelihood. She has no other option excepting to work as domestic worker for her livelihood. She started the work out of her family condition. Her poverty only pushed her to domestic work.

When she started the work as a maid servant she worked with four different houses. In the beginning she received Rs 300 as her wage per month. In two houses she used to clean the floors, washing vessels and cloths, in the remaining houses she used to cook along with the above mentioned works. She never had the habit of taking breakfast in the work places, but sometimes if she felt tired then she used to take food from her owners houses. They never insist her to take food. If there is any food remains in the houses where she worked she used to take that to her home to give the food to her grandchildren. She said that "I have to work at this age as my sons neither they have needed capacity to look after me nor they have concern for me. Even at this age, I have to work and take care of their children also".

For the past 16 years Mahalakshmi has been working in different houses and of them one house is regularly giving this opportunity and paying Rs. 3000 /- per month. Only from that house she used to get Rs. 5000 /- as bonus for Deepawali and in all other houses, she could not get it. She was saying that there is no problem in working in the houses but she has not been treated as a dignified human being. The women folk used to ask her family conditions and this made her to feel awkward. In the houses from young siblings to old men and women never use dignified words. One cannot expect any respect from them and that is why changing the houses frequently. Behaviour of men in the houses where she works is always different which is unacceptable to any women. "Since we rely on their salary for our livelihood, we have to tolerate to a point. If they cross the limit we have to move from that house" she said. "When I was in the fifties, I listened to many comments about me by the men folk in the families where I worked. I never responded and reacted simply because of poverty. But I never allowed them to cross certain boundaries to save my modesty", she observed. Despite her 27 years of her work she could not save any money. She spends everything for her grandchildren. She has lot of physical ailments but she cannot take any leave. She has no money in her saving account. If she takes leave she will be scolded by the house owners and their siblings next day. She never had the habit of taking three meals a day and with only two meals, she is managing herself. She is not aware of the minimum wage act and she confessed that she used to get all the provisions meant for the poor in the schemes of the government of Tamil Nadu. She openly admitted that her conditions would be much worsened if the government of Tamil Nadu has not implemented this much of schemes meant for the poor. She told that she has a plan to join in free old age home when she is not able to work and more freely.

Case No: 2

Mrs. Baby is (name changed) (54) a domestic worker working for a period of eight years. She is married and having two sons and got married. She belonged to BC Vaduga Nayakar community. Her native is Dindigul itself. Her husband never goes for a proper stable job, because he is addict to alcohol and he used to drink daily and quarrel with Baby and beat her too sometimes. At once her hand was fractured because of her husband's beating. She managed her family with the income of her two sons, but when her sons got married her daughter in-laws have not cooperated with her and they wished to lead a nuclear family and hence they went out. Her sons stopped giving any money to Baby and her husband. She could not be able to manage herself without any income. Her husband is also a irresponsible person and he never goes for any job but he want money for liquor and he had been in the habit of harassing his wife. He took all her Jewells including her Mangalsudru (thali) for taking arrack. Her life became miserable. To overcome the problems, she has decided to go for domestic works within the radius of three kilo meters of her residence.

In her working place, the house owner keeps three domestic servants. One servant meant for cooking and another one for cleaning outside the house, sweeping and taking care of the cattles and Baby is for washing clothes and cleaning the vessels. She used to take all the three times food from her owner's house itself. Sometimes she used to take food for home also. Now for the past three years she did not go to her home. She is staying in the owners house itself because her husband keeps on fighting with her and beating her and taking all her salary for drinking liquor. She cannot keep money in her house as her husband takes that money for drinking liquor. When she refuses to give money, he beats her. Her sons are away and they also did not respond to her. Since she has no other alternative, she has decided to stay with the house lord once and for all by abandoning her husband. She is working for her stomach not for money or wealth so she is saying that she doesn't want to work in many houses for financial purpose. Baby is saying that till her death she wants to work for survival and hence, she needs a safer place for stay. Now her house lord gave her a separate small room to stay. Now Baby is having a bank account and also uses to save money. She spends money from her savings if she is unwell. She never had the habit of borrowing money from her owner. The past three years Baby is not attending any social functions and visiting her relatives. Her house lord is taking care of her and she does not want to see her husband and her sons too, she is summarizing all this and said she will work till her death and her saving money will help for her funeral and she does not want anything from her family members.

Case No: 3

Mrs. Mariammal (Name Changed) aged 46, belonging to SC community. She studied up to fifth standard, and her husband working as a coolie and her son is studying twelfth standard. Her native place is Anna Nagar a village near to Gandhigram and now she is residing in the same village. Mariammal working as a servant for the past 20 years and her nature of work is cleaning. She is doing cleaning work for daily wage and for monthly salary with a bonus during Deepavali festival. She gets allowance also. The House owner fixed the payment and monthly four days she gets leave. She did not avail any Medical Insurance, but she has Aadhaar Card, Ration Card and Bank Account also.

She is working in two houses. Sometimes she takes food in the house where she is working but they are not allowed to take food to her home. Mariammal has some amount of saving. She saves 500 rupees for social function per month and she saves 1000 for cultural and religious functions. She spends 200 rupees for medical expenses every month. She faces problems only in financial front. She did not experience any abuse by the house lord at the work place besides she did not find any risk in the activities during working in the house. She borrowed some money as advance from the house lords and she borrowed loan also from the house owner. She has been working more than ten years in the present house. She is a member of Self-Help Group (SHG) organised and promoted by a non-profit organisation located very near to her residence. Besides she is availing all the government benefits like TV, Mixie, Grinder, and Fan from the Government of Tamil Nadu by using her ration card.

She has no own house and she is living in a thatched house. She has some household articles in her house such as TV, Grinder, Mixie, Chair and Cell phone. Regularly three times she takes food (Morning, Afternoon and Night) and sometimes she takes snacks in the evening. She has not undergone any training and she is not willing to undergo any training also. If any institution renders such trainings, she has no time and not interested to participate on Sundays also. She has bank account, ATM card besides she has also debt at present. She borrowed money for her son's education.

She demands for wage hike by request, telling the reason that in fact she is not aware of minimum wage act. The house owner is highly satisfied by her performance that is why she is working more than ten years in the same house. Her house owner is a good individual. For the past ten years she doesn't have any problem with the house owner, and hereafter also. There is no rift between her and her house owner with regard to payment and fixation of wages. She has no idea about when she is going to pay back her debts. Sometimes she felt that she is a slave in the hands of the house lords.

Case No: 4

Jayalakshmi (26) (Name Changed) with fourth standard educational background got married at the age of 19. Her husband working in a small company which is manufacturing steel cots for smaller hospitals with a salary of Rs. 6000 /- per month. They are living in a rented house. Now through her effort the house has been taken for lease for ten years by paying one lakh rupees. She has one daughter and one son and they are in elementary school. Both are studying in a private Christian school by paying regularly. She is working in five houses and earning Rs. 9000 /- per month. She gets food from the houses where she works and feeds her husband, son and daughter in the evening. During night, she is ironing dresses collected from different houses. Since, she belongs to washer man family she knows how to wash cloths and how to press them. Apart from the above job, she used to stitch old clothes. Totally every month she earns around Rs. 12000 /-. She saves money and looks after the family and educates her children. Her husband is always drinking after his work every day. When he crosses the limit, he used to torture her. She has been abused verbally and physically. Rarely he used to give money for the family. She is tolerating all his tortures only because of her children. Since they are studying well getting high score in their studies she enjoys life and she is able to manage the family. She availed all the schemes of the state government. She has 40 grams of gold chain purchased out of her savings. Since she is working in five houses during day time and do other works in her house during night to earn more, she is tired and hence she has to take leave. She has to face lot of verbal abuses from the houses where she is working. She used to tolerate all the verbal scolds because they are paying regularly. In all the houses, the house owners have evinced keen interest in the studies of her children. Her only problem is her husband and in-laws. When she refuses to give money for her husband to purchase drinks, she is beaten up by him. Till he sleeps, she used to work otherwise he will give all kinds of torture to her in the house. She gets support only from her SHG members and the women folk from the houses where she is working.

Case No: 5

Mrs. Vsanthi (name changed) (54) a repatriate from Sri Lanka settled in a village setting. She nurtured hope that she is going to live with her kith and kid. She was married to her close relative at the age of 20. After delivering a male baby she has decided to undergo sterilization as it is the only way by which she can escape from his husband's tortures. Later she came to know that her husband will be of no use as she wants money through her. She has decided to go for domestic work and earn her livelihood. Since she was brought up in Sri Lanka, she has been well oriented in cleanliness, sanitation and hygiene. She started working with two houses initially. A small plot was given by the government for constructing a house. She nurtured an ambition in life that she has to lead a dignified decent life by educating her son. She worked hard and learnt cooking practices in the initial stages in the houses where she worked. Quickly she learnt the skill and increased her earning through cooking food. Since her husband troubled her and abused her by suspicion she made it very clear that she do not want to live with him. Her husband went to Chennai. Thereafter no link between her and her husband. She joined in a SHG and availed loan to construct her house. Wthin five years she has constructed a small house and purchased jewels worth about one lakh rupees. Her son got a post graduate degree. She has availed all the government schemes. Her house is electrified. She has all household equipments. She is a model for many domestic workers. She cooks, clean the house, and washes the dishes in the houses where she works. She earns Rs. 20000 /- per month. Her saving culture is a model for many. But her only problem is her husband. She used to help many to fix maid servants and help to settle dispute between domestic workers and house owners. She helped both. By this kind of activities she has earned a good name. She is also not aware of the act. But she strongly feels that by professionally working in the houses, one can make herself inevitable maid servant. If any maid servant is having capacity and skill, she can easily bargain with the house owners. This is the strategy she followed to enhance her salary.

Case No: 6

Eswari (62), (name changed) is working in two houses and earning Rs. 4000 /- per month she belongs to SC community working in caste Hindu houses. She does all the works from toilet cleaning to dishes washing including cleaning inside the house and washing clothes. She has been working in the two houses for a long. She did that work in five houses before. Now she is not able to do that job as she is old. She divorced her husband ten years ago as her daughter and son advised her to do so. She was a habitual drunkard and disturbed her throughout life by abusing her physically, mentally and sexually. Hence, she divorced. She is living in a small house purchased from her salary. Her daughter educated up to 12th standard and got married to a caste Hindu boy working as a driver in a private company in the same town. Her son studied up to 12th standard working in a company along with her son-in-law. Since my son is also earning, with the salary he earns from the company, her house has been expanded with more facilities and amenities. Both her son and daughter are supporting her morally she leads a peaceful life only now. For about thirty years she led a life without peace. All her earnings have been wasted by her husband. Only after her son and daughter went for a job, she was able to lead a decent life. She nurtured them in a better environment. Both of them strongly advised their mother to send their father out from their house. In such a way, he has tortured her and the family. No relative or neighbor helped her when she faces problems because of her husband. She never faced any problem in the families she worked. Everyone was concerned about her and family. She worked hard and not aware of any act and she bargained with the house owners to fix her wage by telling her conditions. Everyone was cooperative towards her. They are all sympathetic to her and the kids. Now she has a house owned by her. It has all facilities including toilet and electricity. She is arranging marriage for her son. Once it is over she feels that her life responsibility is over. Since there is a demand for domestic workers, no one will disturb the domestic workers. She never faced problems from the families where she worked. She remembered the nature of help they have rendered to her family and especially for her son and daughter. She availed all the government schemes and they are very helpful to save money from her earnings.

Case No: 7

Nayaki (Name Changed) (63), working in three houses. Since she belonged to washer man community she used to help her husband in washing clothes collected from number of families in the place where she lives. Both she and her husband through their hard work earned some amount and saved it for future. They have one son and one daughter. Both have been educated up to 12th standard. Her daughter got married. Her son also got married. Her son is working as driver in a private soda factory. Her daughter-in-law is working as a domestic worker in five families. With the help of the family where she works she got the government assistance to build her house in her plot. Under Green House scheme of J. Jayalalitha she constructed the house. Since she and her husband has got savings, she was able to complete the house constructing with extended facilities. She has got two granddaughters and two grandsons. Both she and her husband developed an aspiration to educate all their grandchildren. Hence, all the grandchildren are studying in private schools. They spent money for their education. It is a joint family. She is managing the entire house. Her son is earning Rs. 7000 /- per month. Her daughter-in-law is earning Rs. 5000 /- per month and she earns Rs. 2000 /- per month. Her husband earns Rs. 10000/- by washing clothes and pressing the washed clothes. They save money through chit funds. They have bank account to get subsidy money from the government schemes. She is regularly purchasing gold to her granddaughter. She informed that she availed all government schemes that helped her to save more money. The families where she and her daughter-in-law work are good families. They provide food for both apart from the salary. Whenever they need money she used to get it from the families where she works. The problem she faces to-day is not from the families where she works. She faces problem because her son. He becomes a drunkard. She takes lot of efforts to correct him. Yet, he is not able to come up fully from the habit. The entire family faces problem because of his behavior. Beyond it, it is a model family for the poor to learn how to come up in life one can learn a lot from this family. She used to argue that she developed her family only by availing all the pro-poor government schemes and through savings. Savings helped her family to come out from poverty and to achieve prosperity. She and her husband maintained good relationship with the families where they work and got all benefits and support apart from wages. She said that their hard work richly helped the family to achieve this much development.

CONCLUSION

It is obvious that the domestic workers have been pushed to this job only by poverty but because of the pro-poor policies and programmes of the state government, their living conditions are far better than poor living in many other states. Though these workers are in SHG, they are not aware of the Minimum Wage Act that regulates their working conditions and wages. Their works are not professionalised. No one has got any formal training to do this work. But their conditions, nature of work, wage are depending on the nature of relationship between them and house owners. They continue their jobs now not because of the poverty alone. It prompted them to work continuously because they nurtured development aspirations and more particularly to invest in the education of their children or their grandchildren. Everyone has got determination and commitment towards a goal and mostly educating their wards. They are not free from problems. They also face problems in the work front but that number is not so high. They enhance their wages not through organizational agitations and negotiations. By using the labour market demand they bargain with the house owners and increase their salary. Only a few are getting frequent advance from the house owners. Others get monthly salary and save money mostly through chit funds. They spent huge money to educate their children through private schools. Not even a single respondent keep her wards in the house. All the respondents are having high development aspirations. They avail all the schemes and programmes of the government meant for the poor. They have bank accounts and they never keep money in the bank account as it will not fetch interest. Their living conditions are improving and they tried to lead a decent and dignified life. They are neither aware of the act nor interested in organizing themselves as they do not have time. They have their own political affiliations. Yet, a small number of respondents have problems in the work front and in their own families. Mostly the respondents faced problems because of their husbands and in laws and sometimes because of their sons. Two things are very clear from their narrations. One that they have strong saving culture. Second, they have strong development aspiration. Mindlessly they invest in their children's education with a hope that their life will be better and bright.

WHAT COMES OUT FROM NARRATOLOGY?

The poor women have worked hard and earned money by facing lot of problems. But they are not able to lead life happily and peacefully. Government of Tamil Nadu has been implementing an array of pro-poor schemes and by using them all poor can come out from poverty. It is unfortunate the males in the family, whether they are husbands or sons, they are the problems for the women. Most of their earnings are wasted only by them. These women continue to work hard to educate their children. They realize the value of education. But it becomes costly as they are educating their kids in private schools. Since public schools are not functioning efficiently to educate the kids well they prefer private schools. If public schools are maintained and worked well they will save more money. Almost all have unanimously said that they need more information about the acts related to their wages and working conditions. If they get small training from organizations and institutions on housekeeping, cooking, old age care, their capacity to earn will be enhanced. They want such organizations.

APPENDIX

Table No. 1 Social Status

S.No.	Category	No. of Respondents	Percentage (%)
1.	OBC	25	52
2.	SC	23	48
3.	FC	0	0
	Total	**48**	**100**

Source: *Computed*

Table No. 2 House details

S. No.	Details	No. of Respondents	Percentage (%)
1.	Own house	20	42
2.	Rented house	28	58
	Total	**48**	**100**

Source: *Computed*

Table No. 3 Type of House

S. No.	Type of house	No. of Respondents	Percentage (%)
1.	RCC	15	31
2.	Tiled	20	42
3.	Hut	13	27
	Total	**48**	**100**

Source: *Computed*

Table No. 4 Sanitation toilet facility

S. No.	Facility	No. of Respondents	Percentage (%)
1.	Individual	15	31
2.	Group	15	31
3.	None	18	38
	Total	**48**	**100**

Source: *Computed*

Table No. 5 Water supply

S. No.	Water supply	No. of Respondents	Percentage (%)
1.	Individual	15	31
2.	Common	33	69
	Total	**48**	**100**

Source: *Computed*

Table No. 6 Drinking water supply

S. No.	Drinking water	No. of Respondents	Percentage (%)
1.	Payment basis	15	31
2.	Public tape	33	69
	Total	**48**	**100**

Source: *Computed*

Table No: 7 Electricity connection

S. No.	Electricity	No. of Respondents	Percentage (%)
1.	Available	48	100
2.	Not available	0	0
	Total	**48**	**100**

Source: *Computed*

Table No. 8 Paying for electricity

S. No.	Payment for Electricity	No. of Respondents	Percentage (%)
1.	Paying	25	52
2.	Free current	23	48
	Total	**48**	**100**

Source: *Computed*

Table No. 9 Provisions from public distribution

S. No.	Provision of PDF	No. of Respondents	Percentage (%)
1.	Getting	48	100
2.	Not Getting	0	0
	Total	**48**	**100**

Source: *Computed*

Table No. 10 All the materials from the PDS shop

S. No.	Materials	No. of Respondents	Percentage (%)
1.	Yes	48	100
2.	No	0	0
	Total	**48**	**100**

Source: *Computed*

Table No. 11 Bank Account

S. No.	Bank account	No. of Respondents	Percentage (%)
1.	Maintained	48	100
2.	No account	0	0
	Total	**48**	**100**

Source: *Computed*

Table No. 12 On your Name

S. No.	Name on account	No. of Respondents	Percentage (%)
1.	Yes	48	100
2.	No	0	0
	Total	**48**	**100**

Source: *Computed*

Table No. 13 Money in the bank

S. No.	Money in the bank	No. of Respondents	Percentage (%)
1.	Yes available	30	63
2.	Minimum balance	18	37
	Total	**48**	**100**

Source: *Computed*

Table No. 14 How to save money

S. No.	Save money	No. of respondents	Percentage (%)
1	Through Bank	0	0
2	Local chit fund	48	100
	Total	**48**	**100**

Source: *Computed*

Table No. 15 Borrowed money for interest

S. No.	Interest	No. of Respondents	Percentage (%)
1.	Yes	17	35
2.	No	31	65
	Total	**48**	**100**

Source: *Computed*

Table No. 16 Children in the family

S. No.	Children in the family	No. of Respondents	Percentage (%)
1.	1 – 2	43	89
2.	Above	5	11
	Total	**48**	**100**

Source: *Computed*

Table No. 17 Education of their children

S. No.	Children's education	No. of Respondents	Percentage (%)
1.	School	34	71
2.	College	14	29
	Total	**48**	**100**

Source: *Computed*

Table No. 18 Do you have gold?

S. No.	Gold details	No. of Respondents	Percentage (%)
1.	Nil	15	31
2.	Up to 2 grams	25	52
3.	Above 2.1 grams	8	17
	Total	**48**	**100**

Source: *Computed*

Table No. 19 Do you have any other asset?

S. No.	Asset details	No. of Respondents	Percentage (%)
1.	Milch animals	7	15
2.	Bike	4	8
3.	Bi-cycle	37	77
Total		**48**	**100**

Source: *Computed*

Table No. 20 Nature of work

S. No.	Nature of work	No. of Respondents	Percentage (%)
1.	House cleaning – washing dishes	25	52
2.	House cleaning – washing clothes – washing dishes	12	25
3.	House cleaning – washing clothes – washing dishes – washing toilets	8	17
4.	Cooking with house cleaning	3	6
	Total	**48**	**100**

Source: *Computed*

Table No. 21 Wage range

S. No.	Wage rate	No. of Respondents	Percentage (%)
1.	Rs. 500 - 1000	5	10
2.	Rs. 1001 - 2000	8	17
3.	Rs. 2001 - 4000	12	25
4.	Rs. 4000 and above	23	48
	Total	**48**	**100**

Source: *Computed*

Table No. 22 Working details

S. No.	Working details	No. of Respondents	Percentage (%)
1.	Working in single house	10	21
2.	Working in two houses	15	31
3.	Working in three houses	15	31
4.	Four houses	8	17
Total		**48**	**100**

Source: *Computed*

Table No. 23 Working hours

S. No.	Working hours	No. of Respondents	Percentage (%)
1.	2 to 5 hours	14	29
2.	5 to 8 hours	25	52
3.	8 to 10 hours	9	19
	Total	**48**	**100**

Source: *Computed*

Table No. 24 Leave facilities

S. No.	Leave details	No. of Respondents	Percentage (%)
1.	Two days in a month	30	63
2.	Three days in a month	14	29
3.	Four days in a month	4	8
	Total	**48**	**100**

Source: *Computed*

Table No. 25 Verbal abuse

S. No.	Verbal abuse	No. of Respondents	Percentage (%)
1.	Yes	22	46
2.	No	26	54
	Total	**48**	**100**

Source: *Computed*

Table No. 26 House owners concern for maid servants

S. No.	Concern for maid servants	No. of Respondents	Percentage (%)
1.	Yes	34	71
2.	No	14	29
	Total	**48**	**100**

Source: *Computed*

Table No. 27 Motivation for domestic work

S. No.	Motivation for work	No. of Respondents	Percentage (%)
1.	Poverty	32	66
2.	Works for children	10	21
3.	Development aspiration	6	13
Total		**48**	**100**

Source: *Computed*

Table No. 28 Health conditions

S. No.	Health condition	No. of Respondents	Percentage (%)
1.	Normal	12	25
2.	Tired	36	75
	Total	**48**	**100**

Source: *Computed*

Table No. 29 Age Profile

S. No.	Age profile	No. of Respondents	Percentage (%)
1.	20 – 30	15	31
2.	31 - 40	19	40
3.	41 - 50	10	21
4.	50 and above	4	8
	Total	**48**	**100**

Source: *Computed*

Table No. 30 Are you in SHG?

S. No.	SHG	No. of Respondents	Percentage (%)
1.	Yes	48	100
2.	No	0	0
	Total	**48**	**100**

Source: *Computed*

Table No. 31 Are you aware of the act?

S. No.	Aware of the act	No. of Respondents	Percentage (%)
1.	Yes	0	0
2.	No	48	100
	Total	**48**	**100**

Source: *Computed*

Table No. 32 Do you have any association?

S. No.	Aware of association	No. of Respondents	Percentage (%)
1.	Yes	0	0
2.	No	48	100
	Total	**48**	**100**

Source: *Computed*

Table No. 33 Do you find uneasy relations in the family?

S. No.	Relation in the family	No. of Respondents	Percentage (%)
1.	Yes	41	85
2.	No	7	15
	Total	**48**	**100**

Source: *Computed*

Table No. 34 Who is responsible for the uneasy relations?

S. No.	Reason for uneasy relation	No. of Respondents	Percentage (%)
1.	Husband	28	58
2.	In laws	13	27
3.	None	7	15
Total		**48**	**100**

Source: *Computed*

Table No. 35 Do you bargain with the house owners?

S. No.	Bargain with owners	No. of Respondents	Percentage (%)
1.	Yes	45	94
2.	No	3	6
	Total	**48**	**100**

Source: *Computed*

Table No. 36 Do you get advance?

S. No.	Advance	No. of respondents	Percentage (%)
1.	Yes	7	15
2.	No	41	85
	Total	**48**	**100**

Source: *Computed*

Table No. 37 Do you find discrimination in the house?

S. No.	Discrimination in the house	No. of Respondents	Percentage (%)
1.	Yes	15	31
2.	No	33	69
	Total	**48**	**100**

Source: *Computed*

Table No. 38 Are you getting all the benefits from the government?

S. No.	Benefits from the Govt.	No. of Respondents	Percentage (%)
1.	Yes	48	100
2.	No	0	0
	Total	**48**	**100**

Source: *Computed*

Table No. 39 Did you get any training for these works?

S. No.	Training undergone	No. of Respondents	Percentage (%)
1.	Yes	0	0
2.	No	48	100
Total		**48**	**100**

Source: *Computed*

NOTES AND REFERENCES

1. Amit Kundu (2007), Conditions of Work and Rights of Female Domestic Workers of Kolkata" **The Indian Journal of Labour Economics**, 50: (4).
2. Bhatt, Ela (1989), Shramshakti: Report of the National Commission on Self-Employed Woman and Women in the Informal Sector.
3. Coelho Karen, Venkat T and Chandira K (2013), "Housing, Homes and Domestic Work: A Study of Paid Domestic workers from a Resettlement Colony in Chennai" **Economic and Political Weekly**, Vol. XLVIII: NO. 43.
4. George Sonia (2013), "Towards Recognition through Professionalization: Organising Domestic workers in Kerala" **Economic and Political Weekly**, Vol. XLVIII: NO. 43.
5. Government of India (2011), **Final Report of the Task Force on Domestic Workers Realizing Decent Work** New Delhi: Ministry of Labour and Employment.
6. Government of India (1988), **Shram Shakti: Report of the National Commission for Self-Employed and Informal Sector workers**, New Delhi: Ministry of Labour.
7. Hamid Areeba (2006), "Domestic Workers: Harsh, Everyday realities" **Economic and Political Weekly**, Aril 1st.
8. ILO (2012), **Domestic Workers across the World: Global and regional Statistics and the extent of legal protection** Geneva, ILO Publications.

9. Moghc Kivan (2013), "Organising Domestic Workers in Pave City" **Economic and Political Weekly**, Vol. XLVIII: NO. 43.

10. Neetha. N (2008), "Regulating Domestic Work", **Economic and Political Weekly**, 43(37).

11. Neetha, N (2013) "Minimum Wage of Domestic Workers: Mirroring Devalued House Work, **Economic and Political Weekly**, 48 :(43).

12. Sankaran, Kamala (2013), "Domestic Work, Unpaid Work and Wage Rates" **Economic and Political Weekly**, Vol. XLVIII: No. 43.

13. Vadageri, Bhagyashree. S and Soundari M. Hilaria (2016) "Socio-Economic Problems of Domestic Labourers in Karnataka" **Indian Journal of Economics and Development** 4:(4).

Chapter II

A Micro Analysis of the Life and Livelihood of Mayana Vettiyan in Tamil Nadu

INTRODUCTION

"Mayana Vettiyan" is an individual designated by a community among the scheduled caste families to perform the task of burying and burning the dead human bodies and dead animals in the villages of Tamil Nadu. It is an assigned function to the Dalit families by the dominant communities through the practices evolved historically over a period of time. It is being practiced in the rural areas without any resistance. The same practices are in vogue in urban areas also. Yet the conditions with which the practices carried out in the urban areas are somewhat different in terms of the hardships undergone by the Mayana Vettiyans in the rural areas[1]. It is a task can be done by anyone provided it is done scientifically and professionally as it is being done in the western country. It can be made as a skilled work and this work has to be done by a skilled worker. If it is done through skilled worker, payment has to be made as fixed by the government. But assigning the work to a few Dalit families and the work carried out in an unscientific and hazardous way is really inhuman and condemnable in democratic governance arrangement. It was declared abolished in Tamil Nadu through a government order No.74 dated 24th May, 2002. Despite the passage of the government order, this practice is being continued in rural Tamil Nadu. There are many progressive acts in India to facilitate the marginalised and deprived to empower them but they are not implemented letter and spirit[2]. In the same way it is also not implemented in Tamil Nadu. There are innumerable hurdles and barriers in implementing such revolutionary reforms and acts in India. Enacting legislation is not a complicated task. Implementing the enacted legislation is the real task and achievement. It has been seen in independent India, how difficult it is to implement the provisions of the constitution of India for the benefit of the poor[3].

Tracing the history of the practice of burying and burning dead bodies by a few individuals from a community hailing from the lowest strata of the society goes back to epic age. All the activities associated with burning and burying the dead bodies had been performed with varying degrees of rituals with differences from region to region. Rituals have been changed and yet the conditions with which the burial and burning have not been changed. As a result, the persons involved in burning and burying the dead bodies face volley of problems. The problems associated with burying and burning are also equivalent to that of the problem of the manual scavengers[4]. Neither the community nor the policy makers are concerned about the plight of the graveyard workers called 'Mayanavettiyan' in Tamil Nadu. Their pain and agony are not understood properly by the policy community. Their problems have not reached the corridors of decision making bodies despite the earnest efforts of some of the civil society organisations. Even if they take decision at the state level to help them to redeem their problems, mostly it is cosmetic and it is not implemented letter and spirit. But in the urban areas, the conditions are somewhat better wherever they are assisted by the civil society organisations[5]. Wherever the civil society organisations have involved in helping the graveyard workers, the conditions of the work have been improved as they follow some scientific procedures. Civil society organisations help the graveyard activities only in the urban areas. The problems of the graveyard workers in the rural areas have not reached strongly the vicinity of the civil society organisations. Only a few civil society organisations are involved in the rehabilitation works of the graveyard workers. Organizing the manual scavengers, graveyard workers and cobblers are not the easy job as one finds in organizing other social groups. This may be one of the reasons why many civil society organisations have not evinced interest in organizing such marginalised groups. Since the number of people involved in this task is minimum in every village, their plight has not drawn the attention of the policy community. Even though there were ministers hailing from the same community, they have not come openly to support their cause. They are closer to power centre and they know the problem and yet they are not representing the problems of this category of people.

The graveyard workers are not a distinctive social group in the village setting to perform the above tasks of burying and burning the dead corps. They are also individuals from Dalit groups. They also live with other Dalit groups. No way, they are inferior or superior to the other Dalit families. "Vettumai" is a traditional job to be performed by the Vettiyans for the farming communities in the rural areas. Only a few designated families among the Dalits perform this task. This job has been conferred on certain families by the dominant caste groups in the rural areas. This job cannot be done by others. This is a common unwritten agreement among the communities. The individual families who perform the task of burying and burning the dead bodies and the communities in the rural areas on mutual consent carryout the tasks without having any conflict and disagreement even though there is no written agreement. Their operational units are not synchronized with either Gram Panchayat areas or revenue panchayat areas. Their area of operation is determined by the communities. It is also not decided by the communities now.

It is an age long practice followed in the rural areas for several centuries. They are not only burying the dead bodies and burning the same but also removing dead animals from the habitats of the higher caste groups. These activities are going on in Tamil societies over centuries. These jobs are difficult to carryout and yet they are being carried out by the dalit families. In the fast changing modern world, any activity can be carried out by any individual provided they are done professionally with adequate care attention and skill. If any work is professionalized through a formal system, that would attract the attention of everyone as it would help them to earn money. In India many of the manual works are unprofessionalised and they are earmarked for a few sections and they have to carry out all those activities as they are mandated by the community through a system of informal arrangement. 'Vettumai' is one such profession to be done by the dalits in the villages. What are all the problems they face and how they do the job are being investigated in this work.

To study any social problem or issue a well formulated theory is an imperative need. Theory will be a goal post and it will enable the researcher to pursue the process of investigation more scientifically with precision to add value to the existing body of knowledge on this subject. Hence, to study this problem of marginalization the researcher has searched for a meaningful theory. On the search "Basic Grounded Theory" has been considered more suitable for this investigation and hence this theory has been adopted in this study.

BASIC GROUNDED THEORY

Basic grounded theory[6] is qualitative method adopted in social science to map up certain peculiar social realities which are unique and dissimilar to the general issues and problems. This theory helps the researcher to explore new area very deeply and the social phenomena are intricate and delicate which are to be investigated. Every aspect of the social phenomena has to be coded and is to be described, analysed and explained. It is applicable to the unique cases. The researcher can probe through any method and bring to light the realities. This method can be widely used in Indian and African social setting as differences, disparities, derivations, discriminations, and deprivations are abundantly explicit in the social constructions. Each segment of the society has peculiar character. They have the unique characteristics features which cannot be compared with any other in the same society. The grounded theory's canvass is wide and the freedom of the researcher and the integrity of the researcher is crucial[7]. For the investigation, the researchers need not depend on any conventional tool, framework or theory.

METHODOLOGY

To get to know the details, a focus group discussion was arranged. It was arranged in Alathambadi village in Thiruthuraipoondi Taluk in Thiruvarur district on 7th May, 2017. It is part of composite Tanjure district. In this district the left political parties have established their grip as they struggled for the emancipation of the working class in the rural areas and they have succeed in their attempts in considerable measure. Hence, this study was conducted in this area. This study is a preliminary one to launch a major study in Tamil Nadu. An extensive and elaborate discussion was held with a group mostly the vettiyan family members. All are men. Of them three are 'Vettiyans' and others are only helpers. Mr. Radha (55), Gyanasambandam (49) and Thangavel (55) are the vettiyans performing the job since their earlier days. The participants who participated in the discussions had long year of experience and hence they had been chosen as respondents. They are performing this job as per community practices. In the focus group discussion, there were other members but they are not designated individuals for the job. But they perform the task as instructed by the designated individuals. But all the individuals have involved in the task of burying and burning the dead bodies. From the researchers perspective they are helpers and they also know the practices. The discussion went on for four hours. Though the helpers also participated, helpers have not spoken much as the Vettiyans handled the questions directly. Whenever, wherever the helpers are allowed to speak by the Vettiyans, the helpers have responded. They have been informed that the interview was only for research and not for anything.

PERCEPTION, DUTIES AND RESPONSIBILITIES

All the three respondents are "vettiyans" designated by the community as perceived by them. While explaining it in the Focus group discussion, all the three categorically made a statement that they are the responsible persons to perform the task. This was considered as right conferred on them by the community traditionally. No other individual can perform this duty without their permission. They informed that whenever any death happens in any family, the family members immediately inform any of the three about the death. When it is communicated, it is the responsibility of the vettiyan to go to the place immediately and ask for the details of the works to be carried out. To perform their duties and responsibilities they will prepare themselves and their team. They know the ritual pattern of each caste group and accordingly they take steps. They have not perceived that it is a job thrusted on them. They considered that it is a right. They are of the opinion that it is not confined to village panchayat area. Their area of operation is determined by the settlement of few cohesive communities. It is neither village panchayat area nor revenue panchayat area. They cover only community panchayat areas. Both communities and the vettiyans know their area of jurisdiction. Neither the communities nor the vettiyans feel any problem in the task and their relationship.

People from different communities in the villages inform such tasks only to these individuals. These responsible and earmarked Mayana Vettiyans have to decide how many have to attend, how to attend and what could be done. Twenty years back number of activities have been carried out by the vettiyans from the day of burning to the 16th day of burning rituals. They had been carried out by the vettiyans on consent as observed by the participants. No doubt they have faced volley of problems in performing these functions. But they never feel that they have to give it up. They feel that they are also doing a function as Brahmins do it in the temples. But now they have started realizing the problems they faced are imposed by the traditions of the communities. Yet, they have not prepared for resisting it or to abandon it. Now they compare themselves with Brahmins. Why Brahmins are respected why the vettiyans are not respected? Why people pay what Brahmins demand? Why vettiyans have to accept what the community gives for the job they do? are the questions now they raise among themselves.

They perform variety of responsibilities related to death and death related events. They carry out both burials and burning the dead bodies. To perform the above activities, and for which there is no fixed amount earmarked either by the community or the vettiyans. When death occurs in the affluent families they get more wages than their average wage they earn per day. They cannot expect much money from the poor families to perform the same works. Depending on the requirements, workers are being mobilized and for which vettiyan is responsible. Vettiyan is being contacted if any death happens. It is his responsibility to mobilize workers to perform either burial or burning the dead bodies in the villages. The work starts from informing their relatives about the death of a person, drum beating, preparing the burial ground or burning ground for burial or burning. From information given to the relatives of the deceased to burning the dead bodies, all the activities are to be performed by the vettiyans. There is no fixed cost for each of the work. The expenditure and wages are highly arbitrary and the vettiyans have no say and they cannot demand. At present barring burial and burning all other activities are being carried out by many. Technology has come to the rescue of the dalits who have been involved in this burial and burning work. Technology has come to all other activities barring burning and burial in the rural areas. If it is burial, it is only an hour work. If it is burning, it is almost twelve hours work after laying the body on the burning place. Mostly that work is carried out during the night. On an average one individual can get Rs. 500/- per night and Rs. 350/- for assisting the person who is involved in burning the dead bodies.

They never take any safety measures while performing their job of burning the bodies. Even they never cover their legs by using shoes. With only a knife and a small stick they manage the work. Before they start the work, they take alcohol. They perform the task in the influence of alcohol. Bodies are different types. No one inform them about the cause of death and with great risk they do the work. Only during the burning they understand the nature of the body. Mostly they will be in a drunken mood. They will be in the job to complete the target. It will take seven to twelve hours. Till the process of burning is completed the vettiyans have to be in the place in the midst of fumes. Each community has its own place either for burial or for burning. In the rural areas, the burial and burning ground has been used by the public for open air defecation. So cleaning the place is the difficult task. It is just like a manual scavenger vettiyan has to clean the place. Once the burning is completed, it is to be informed in the respective families.

After burning the body, they have to keep the burnt body in such a way that the family members have to easily trace the skull, leg and hands for the ceremony they do in the next day. While burning the bodies, they notice changes and by noticing it they have to procure fuel and add it in the fire to complete the burning quickly. Some of the bodies produce intolerable foul smell and the vettiyans have to tolerate the smell till the job is over. Till the burning is over the vettiyans will be inside the smoke. Dead bodies are different types. Suicide committed bodies and HIV infected bodies are also burnt by them. The vettiyans do not know the details of the death or causes for the death but they have to carry out the job as per the decision of the community. Some of the intricacies of the death of some individuals came to light only through opinion makers in the village. The vettiyans can guess from their body reactions while in burning about the nature of death.

Sometimes bodies come due to suicides, and in such occasion, the bodies have to be burnt quickly. During that time heavy fuel has to be added. Normally white sugar used to be poured on the body. Fumes come out of body while burning it. They cannot move out from the fumes. They smell it and it is a difficult process. The odour of the fumes vary from body to body. They cannot move away from the site. They have to enable the body to burn properly. Hence, they will be very closer to the site. Only instrument they use during the burning process is a long wooden stick with one lengthy knife. In normal condition, they cannot tolerate the smell of the fumes and that is why they consume alcohol. Everyone knows that the people involved in burning dead bodies consume alcohol. They reported that for consume alcohol, they liberally get money from the community. After burning dead bodies, they cannot do any work for two or three days. They will be so tired or sometime they are not in any mood to work. So the next two or three days they cannot earn anything. But it is not compensated, they reported. In many of the occasions, they fell ill due to unknown reason. They have to spend money for the treatment. No one is giving money for the treatment. Though women are involved in this task[8], in East Tanjure district women are not involved in this work. Women involved in this job are not new. It is in practice in other states also[9].

There is yet another job the vettiyan has to do. Removing dead animals' mostly dead cows and bulls. Vettiyan receive information about the death of animals in the farmer's house. Community gives the information to the vettiyan through some one about the death of cows or bulls. It is the role of vettiyans to remove the dead animals. Vettiyan has to organise four members to carry the dead animals to their habitat. They have a specific place to remove the fur from the animal. After removing the skin, the animals have to be buried in a safe place. The whole process takes minimum of four to five hours time. During summer burying the dead animal is a difficult one as taking pit for burial will take longer time. For the whole process nothing is paid. There are some individuals having realized the labour lost in this process, they give money. Others feel that it is the duty of vettiyan to remove the dead animals and hence they will not give money. The skin of the animal can be sold. But the cost of the skin will not meet the cost of labour involved in disposing of the dead animals. At present, this problem has been reduced as animal population is dwindling down and drastically reduced. Further, the farmers feel by burying the dead animals in their plantation will become good manure to the plants. Hence, they bury the dead animals in their plantation itself.

After the present generation, there will not be anyone willingly to do the work anymore. This vettiyan groups have internalized this work as their duty and commitment to the community and hence they are doing it. The sons and daughters are in the schools and they do not like this job to be carried out by their fathers. Hence, there is no guarantee that their wards will continue this practice in these villages after their death. By going through the nature of the task, the vettiyans can be paid more. But normally it does not take place. Almost all of them had the feeling that they are not paid well. Yet they felt that they should not bargain with the community since it is related to death. They feel that it is beneath their dignity. The community has to realize that they are paying less to vettiyans for the work done by them. Previously, there was a practice that vettiyans have to collect paddy during harvest seasons by visiting paddy field. By collecting paddy, every household of vettiyans would collect paddy to the requirements for the family for a year. Now for the past two years, there is no cultivation of paddy during farming seasons. While interacting with vettiyans one will find their generosity. They stated that it is unfair to bargain on the dead bodies. But it is the responsibility of the community to look at the wage part of it. The same vettiyans said that they are fighting for their wages when they involved in agriculture work.

In the event of vettiyan sick for a particular day, he has to nominate a person among his community to perform the functions. Even during sick, they are discharging their duties. But nobody to bother about their health. While interacting with them, on their health conditions, many indicated that they have multiple health problems. One will find many scars in their body. When they are asked about it, they informed that they got several wounds during the process of burning the bodies. Many have problems in their eyes. They have several ailments and for which they get treatments in the hospitals. They consume medicine worth about Rs. 1000 per month. Lot of money has been spent on health. Most of the occasions, they borrow money for interest not in the bank but from the money lenders. Of them no one has got bank account and no one has got savings. No one expressed hope that their life can be changed or transformed. Their life is to earn for their food and other daily requirements. They are always in debt. Savings is to repay the loan which they borrow for their family needs. They have no answer to the question, why they are not refusing to do the obnoxious works? They feel that it is the work imposed on them and yet they feel that they have to do the work as it is destined.

There is an organisation working for this category of people. It is called "Mayana Vettiyan Sangam"[10]. It was established in the year 2007 by the Tamil Nadu Labour Union at Madurai. It wanted to bring all the vettiyans all over the state under one fold. After the enrolment of the member, the organization conducted a conference on 7th January, 2008 at Madurai. When they organized the conference a Minister Mrs. Tamilarasi Ravikumar for Adi Dravidar Welfare, Government of Tamil Nadu participated. The organization has put up several demands to the Government of Tamil Nadu. They are: creation of separate board for them; minimum salary for them, old age pension, maternity benefits for the women workers; rest rooms and toilet facilities in burning ground. These are all the major demands projected by them and handed over to the Minister. They continuously organized some rallies to press their demands subsequently. But the association seems to be a closed chapter as it could not mobilize vettiyans to build a movement. No civil society is engaged in mobilizing them. But the respondents are not aware of the association and its activities. Everyone wants assistance to protect their health. They want a specific scheme to ensure their livelihood and help to protect their health.

Box - I

Vettiyan becomes the President of Gram Panchayat

Koduvilarpatti Gram Panchayat president post has been reserved for SC and hence Mr. Chinnavaradhan has been asked to file his nomination. He is a vettiyan performing his traditional role of burning and burying the dead bodies. He promised the people even after the election he will continue the work of burning and burying the dead bodies. He won the election and continued the traditional duties and responsibilities. Many persuaded him to do the vettiyan work. But he did the job as promised to the people. He mentioned that he need not explain the problems of vettiyan as everyone knows. What he informed is that he brought several facilities to graveyard. No one has taken such steps to beautify the graveyard. It is a neglected place. Everyone comes to graveyard but their stay is for a few minutes. Hence, it is not registering in their minds. But I stay here for the whole night when I burn dead bodies. Hence, the condition of the graveyard is in my memory. When I got an opportunity I have started providing all the facilities. Water facilities, shed for burning dead bodies, planting of trees, approach road to graveyard have been provided. He also said that he maintained cleanliness in the graveyard. A compound has been constructed for the graveyard. He said that he performed his task as president of the Gram Panchayat along with his traditional role.

— Dinamalar, 4.7.2017

Box - II

Women in Graveyard

Even men dare to visit graveyard during the time of burning the dead bodies. But Ms. Vairamani is performing the role of a vettiyan in Chokkampudur graveyard in Coimbatore district, Tamil Nadu. She performs her duty mostly in night without having any support of man. It was the duty of her father to perform burning and burying dead bodies as per the practice of the community. She got married and has three children. She married at the young age as it was decided by her father. Husband could not take care of her family through his earning. At this juncture, her father died and nobody to take responsibility to perform her father's role. Hence, she has decided to take up this job as she can earn through this work. Since she helped her father to perform this job, she has no fear. She considered it as a divine place. She fixes Rs. 2000 per body either for burial or for burning. She keeps one assistant to do the job. She gets this amount after a tough bargaining. Many of the occasions, dead bodies are brought to burial ground without relatives. During those occasions she cannot ask for labour charge. Her children are studying in the schools. Normally, six to seven hours the dead bodies will take to be consigned. The work is tiresome and it has to be performed in smoke. Dead bodies are brought to the graveyard but no one inform the cause of the death. Only during the time of burning the dead bodies, one can guess the reasons for the death. It is not a healthy job and she had no other option, she had decided to do.

— Dinamalar, 9th June, 2013

Box - III

Mayana Vettiyan's Green Route

Everyone sympathize with graveyard workers when one visits graveyard and forget about them after the departure from the graveyard. Their conditions are deplorable. They work in hazardous conditions by risking his life. Everyone knows yet no help from anyone. Ministers, people representatives, business man, intellectuals, officials are visiting graveyard and they will be there briefly but no one take any step either in cleaning and beautifying the graveyard as in the west or rehabilitating the graveyard worker. But Mr. Hari, a graveyard worker in the graveyard at Thathaneri cremation ground in Madurai has done a yeoman service to the society. Apart from burying and cremating the dead bodies, he planted saplings in the cremation yard and watering them continuously for the plants to make it a green cover. No one gives money to buy saplings. He pays from his pocket. When he was asked, he said that he has a concern for the people who visit graveyard along with dead bodies as they are in scorching sun or in rain without having any shade in the graveyard when they come along with dead bodies. "They are not bothered about me or themselves. But I have a concern for them" he said while interacting. He further lamented that people spent much money for temples but they are not interested in keeping the cremation ground clean, neat and green. Sympathy, empathy, concern and compassion are with the poorest of the poor not with elites and educated.

— The New Indian Express, 24th September, 2014

CONCLUSION

Vettiyans have to be relieved from such hazardous works and they have to be rehabilitated. The Government Order passed by the Government of Tamil Nadu has to be implemented letter and spirit. Till government finds alternative to this practice of burning and burying the dead bodies, the grievances of the Vettiyans have to be redressed. Protective measures have to be taken by giving instruments and safety equipments for the works they do. The wage has to be hiked by doubling the wage of unskilled workers. Minimum wage has to be fixed for the work and that can be done by the government. Burial and burning ground should have some basic facilities and the place should be maintained as clean as the divine place. An electric light provision has to be made in the burning ground. Water and toilet facilities have to be created there. Shade giving trees have to be planted in the premises of the burial and burning ground. The graveyard has to be maintained neatly and cleanly. Compound walls have to be constructed. All the above facilities have to be created by the Gram Panchayat. Professional training has to be conducted for them to perform the task more scientifically. There are modern cemeteries constructed by various civil society organisations. These cemeteries never use firewood to burn the bodies. After performing the religious rites, the dead bodies will be inserted into the cemetery. The body will be burnt in the natural heat generated in the cemetery. Government of Tamil Nadu can think of aliternative mode of disposal of dead bodies as devised by civil society organisations. Considering the quantum of firewood used for this purpose, an alternative mode can be thought of. Thus for every five panchayats one cemetery could be created and they can be managed professionally by the local bodies. User fee can be levied from the users. Cost of maintenance can be met from the user fee collected from the users. Thus this vettiyans can be liberated from this unscientific task and modernize the very process of burning the dead bodies.

NOTES AND REFERENCES

1. Vidya Venkat "They deal with death every day at work" **The Hindu** 24th August, 2008.
2. Bhasha Singh, **Unseen: The truth about Indian's Manual Scavengers**. New Delhi: Penguin Books, 2014: B.N. Srivastava, **Manual Scavenging in India: A Disgrace to the Country**. New Delhi: Concept Publishing Company, 1997.
3. Grenville Austin, **Working a Democratic Constitution: A History of the Indian Experience** New Delhi: Oxford University Press, 2012.
4. D. Senthil Natarajan "Two women oversee burial ground work" **Deccan Chronicle** 25th August, 2016.
5. Kathy Charmaz "The Grounded Theory Method: An Explication and Interpretation" in **Contemporary Field Research** (eds) R. M. Emerson. Boston: Little Brown, 1983.
6. U. Flick**, An Introduction to Qualitative Research** London: Sage Publications, 1998.
7. "Burning the body in the night by women vettiyan" (in Tamil) **Dinamalar** Tamil Daily, 9th June, 2013.
8. B. Madhu Gopal, "Women caretaker at burial ground" **The Hindu** 19th March, 2014.
9. Tamil Nadu Mayana Vettiyan Sangam, **Epic Age to the Present: The Life of the Mayana Vettiyans A Report** Madurai: Tamil Nadu Labour Union, 2009.
10. Gokul Vannan, "Grave Digger Takes Green Route to Mark Amavasya at Madurai Cremation Ground" **The New Indian Express** 24th September, 2014.
11. "A Women in Challenging task at Graveyard" **Social** 8th March, 2017.

Chapter III

A Micro Analysis of the Process of Implementation of Manual Scavengers Act 2013 in Tamil Nadu

INTRODUCTION

Sanitation and scavenging in India are inseparable and both are in deplorable conditions. Even after seventy years of democratic governance and implementing plethora of development schemes, neither the conditions of sanitation nor the conditions of the scavengers have been improved as expected by founding fathers and mothers of India. In India 59% of the people used the open space either in city or in the villages for defecation. It is an ample proof that our people's life has not been guided by science (Gita, Ramaswamy, 2005). It causes serious health problems to the people. Reports produced by UNICEF highlighted the enormity and seriousness of the health problems in India. It is a reality that neither the state authorities, nor the civil society and science community responded with the intensity of seriousness expected from them. The above conditions have reflected the insensitivity of the political communities, policy community, bureaucracy, the scientist, the civil society organisations and the media in India. Is 70 years not enough to achieve sanitation by adopting science and technology? A war ravaged country like Ruwanda is able to achieve tremendous success in sanitation, why India is not able to achieve results in sanitation project despite its achievements in economic growth and creating scientific man power. It was the basic work started in the National Building process. M.K. Gandhi had given much thought to it and found out ways and means to tackle it. He gave targets and processes for sanitation work to the constructive workers with the thinking and hope that it would become a massive work to be launched by the Gandhian workers and Congress party workers as the Indian National Congress party claimed the legacy of Gandhi. Still we teach constructive programme of Mahatma Gandhi to the students in the classroom while teaching Gandhian thought to the students. Even the Gandhian Institutions and Departments have not moved beyond teaching. On the issues of sanitation and liberating manual scavengers, the two builders of modern Indian society M.K.Gandhi and Dr.B.R. Ambedkar differed sharply as on other issues (Bhasha Singh, 2012). M. K. Gandhi wanted to change the condition through mass participation in this task with the hope that the participation of people in sanitation work will bring a total transformation in the thought process of people. Gandhi's concept of participation is meant for attitudinal and behavioural transformation of the participants. In the absence of scientific management system of sanitation, it is the only way to change the attitude and behaviour of the people through their participation in the sanitation work, cleanliness could be maintained in the society. Hence, Gandhi wanted to attend to the sanitation work immediately and more seriously. That is why he has accorded more importance to sanitation than freedom and liberty.

Since independence targets have been fixed for construction of toilets and thereby a perception has been created that sanitation issues can be addressed systematically through the construction of toilets by the initiatives of the government. Mere construction of toilets will not solve the problems of sanitation and scavengers. Even today we have the same approach what we had in the past. In "Swatch Bharat Mission" Indian government wants to construct 12 crore toilets in 2019. To achieve the goals of swatch Bharath, construction of toilets is alone not enough. It needs a toilet culture in the society. Solutions can be found out through appropriate science and technology and changing the attitude and behaviour of the people. It requires money, system, approach and change agents. Unfortunately in India the conditions of the manual scavengers have been linked with the issues of caste system (Bezawada Wilson, 2016). Thus, both issues have not received the needed attention and seriousness among the policy community and the scientist even till today. As a result, a microscopic minority involved in the struggle for liberating the manual scavengers from the inhuman and dangerous activities. Since the numerical strength of manual scavengers has not reached the critical mass the issues of the scavengers could not draw the attention of the political class also. Beyond that they are also not conscious about their own living conditions.

The struggle for liberation of manual scavengers is a long history. It's origin from the caste hierarchy reveals the age old practice of the manual scavenging in India. Practices of manual scavenging vary from state to state and thereby conditions of the manual scavengers also vary across the states. Organisations working for rehabilitation and their intensity of activities vary from state to state. Equally the state government's response to the demands of these organisations also varies and yet the conditions of the manual scavengers are deplorable (Bhasha Singh, 2012). It warrants the policy community to act fastly to save the name of India from the community of nations. Organizing them for a movement is also not so easy as their struggle for livelihood is intensive and highly complicated. Their standard and quality of life are far below the decent and dignified human life. They live in sub-human living conditions. After prolonged struggle both political and legal the manual scavengers got a relief through a new act called Employment of Manual Scavengers and Dry Latrine Prohibition Act 1993. It is to be noted here that these practices are inconsistent with the right to live with dignity which is an essence of the fundamental rights guaranteed in part III of the Constitution. Their life is always at risk. One cannot argue that the Government of India did nothing for them. it also initiated various measures to transform them (B.N.Srivastava, 1997)

Despite the provisions in the constitution of India, enactment of Employment of Manual Scavengers and Construction of Dry Latrine (Prohibition) Act 1993, the conditions of them have not been improved (Government of India, 1995) instead their struggle has been intensified because of the tardy approach of the state governments in implementing the act fairly (Vimal Kumar, 2012). The manual scavengers struggle continued along with legal course they adopted to make the state governments to operationalise the new act (Siddaramu, 2013). The system and the process continued in sanitation work as usual without any course correction by indicating the absence of clarity in definition and operationalization (Susan E. Chaplin, 1997). Finally through a legal battle, under the direction of the Supreme Court totally a new act has been enacted in the year 2013 (Government of India, 2013). The prohibition of employment as manual scavengers and their rehabilitation act 2013 was passed in Indian parliament. The new act sets out its objectives clearly. They are: a, identify and eliminate the insanitary latrines and; b, prohibit employment as manual scavengers and hazardous manual cleaning of sewer and septic tanks. Even after this passage of the act the implementation of the act is weak and one has to seek the remedy only from judiciary for proper implementation. Althrough arguments have been projected that imprecise definitions without having much clarity in operationalising the provisions of the act by the policy community and bureaucracy, the new act has come with precise definition. It is to be noted here that when the struggle was on, the policy community, bureaucrats and the scientist could have found out an alternative system and process to maintain sanitation scientifically and to liberate the manual scavengers successfully. But the reality reflects the ground zero. Deaths have been reported due to the age old practice of using manual scavengers to clear the blocks in sewerage system. Practice of disowning them also continues because of the new Act. International organisations such as the international labour organisations unequivocally classified the three forms of manual scavenging are in practice in India. They are involved in removal of human excreta from public streets and dry latrine, cleaning septic tanks and cleaning gatters and sewers. The 2011 census data reveal that 180657 households involved in manual scavenging as a profession for their livelihood.

Tamil Nadu is a land of E.V.R. Periyar and known for its pro poor policies and programmes. It is the most fast urbanizing state. In all parameters, it is very near to Kerala in human development indicators. Yet the state has more number of manual scavengers. Equally condition of sanitation is not impressive. Now only momentum has been picked up and village panchayats and municipalities are declared open defecation free areas. It has not reached even 50% of total Tamil Nadu. There are three communities namely Pallars, Parayars and Arunthathiyar involved in inhuman practices of manual scavenging. The struggle to liberate the manual scavengers in Tamil Nadu is intensive and hard despite the Governments pro poor approach. It is to be noted that along with other organisations, the Director Change India organisation through the judiciary is relentless fighting for the cause of manual scavengers till today even for implementation of the act and for the compensation to the victims of manual scavenging. On this above issue the Government of Tamil Nadu has not shown the seriousness as it exhibits on the other issues of the poor. It is a paradox in Tamil Nadu. This riddle prompted us to undertake a study as to what extent the state progressed in rehabilitating the manual scavengers as per the directions of the Supreme Court and the Act. Before moving into the analysis of the process of implementation of the Act, it is necessary to find out the salient features of the Act with a view to look at the Act from the perspective of operationalization.

SALIENT FEATURES OF THE ACT

The prohibition of employment as manual scavengers and their rehabilitation act, 2013 defines manual scavenging very clearly for the benefit of easy operationalization. The earlier Act, 1993 faced many problems only in operationalization. Manual scavenger means a person engaged or employed on regular or frequent basis by an individual or a local authority or a public or private agency for manually cleaning, carrying, disposing of, or otherwise handling in any manner human excreta in an insanitary latrine or in an open drain or pit into which human excreta from insanitary latrines in disposed of or on a railway track before the excreta fully decomposes and the expression "Manual Scavenging" shall be interpreted accordingly.

The salient features of the Manual Scavengers Act, 2013 as follows:

- No person, local authority or any agency shall construct insanitary latrines, engage or employ a manual scavenger.
- Every person engaged or employed as manual scavenger shall be immediately discharged from doing manual scavenging.
- Every insanitary latrine existing shall be demolished or converted into sanitary latrine.
- No person, local authority or any agency shall engage or employ any person for hazardous cleaning of sewer or septic tank after one year of the date of commencement of the Act.
- For employing manual scavenger and or constructing insanitary latrines – first time imprisonment up to one year and penalty of Rs. 50,000/- or both and for subsequent violation imprisonment up to 2 years and penalty up to Rs. 2,00,000/- or both.
- For employing any person for cleaning hazardous cleaning of sewer or septic tank, imprisonment up to two years and fine up to 2,00,000/- or both and for subsequent violation imprisonment up to 3 years and fine up to Rs. 5,00,000/- or both.
- Every offense under the Act is cognizable and non-bailable.
- Offenses under the Act may be tried summarily.
- Executive Magistrate shall trail the offenses as Judicial Magistrate of First Class.

RESPONSIBILITIES OF THE LOCAL AUTHORITY

- Every local authority shall carry out the survey of insanitary latrines in their jurisdiction within two months from 6 December, 2013
- Local authority shall publish the list of insanitary latrines.
- Demolishing such insanitary latrines or convert them in to sanitary latrines within six months from December, 2013.
- Constructing sanitary community latrines as required in nine months from 6 December, 2013.

- To eliminate practice of open defecation, constructing adequate sanitary community latrines within three years from 6th December, 2013.
- Making arrangements for keeping community sanitary latrines hygienic all times.
- Conducting survey for identification of manual scavengers and publishing the list within two months from 6 December, 2013
- If any person is engaged in manual scavenging, he/she shall file application to concerned local authority. And local authority shall inspect within 15 days of such application and shall include in the list.

REHABILITATION FRAMEWORK

- Photo identity card to the manual scavengers will be issued
- Initial cash assistance will be given
- Scholarship to the children of manual scavengers should be provided
- Allotment of residential plot with financial assistance for the construction of the house or a ready built house, with framework of relevant scheme should be in place
- Training in livelihood skill for all the people who are relived from manual scavenging
- Concessional loan with subsidy for taking up alternative occupation has to be arranged

MONITORING MECHANISM

A monitoring mechanism has been created. At the central level the Minister for Social Justice and empowerment and at the state level the Chief Minister are the chairpersons of the monitoring committees. Vigilance committee will function at district and sub-division level under the chairpersonship of the District Collector and sub-divisional magistrate respectively. By keeping the framework of salient features of the act, an attempt is made to evaluate to what extent the main provisions of the act have been implemented in Tamil Nadu. In the past one decade India saw a "Rights Revolution". Right to Information, Right to Employment, Right to Education, Forest Rights Act, Right to Food, Right to Protect street vendors and few others have been passed in Indian Parliament. By passing such revolutionary acts in the context of globalization, liberalization and privatization India has drawn the attention of the world to understand the process by which poor are protected from distress and vulnerability in India. Here it is to be seen that all these acts have to be translated into action only by the state and local governments at the ground. Each state is functioning in a distinct socio economic and political culture despite the common framework of governance and administration (Samuel Satyaseelan, 2013). Tamil Nadu is one of the progressive states in India in Industrialization and social development activities. It is a fast urbanizing state. It is a model in evolving pro poor schemes. Despite all progressive indicators, it could not achieve sanitation on par with other indicators of development. It is a riddle for many of the development thinkers and practioners of development. Against this background an investigation is undertaken in the field.

OBJECTIVE OF THE STUDY

1. To analyze the life and livelihood conditions of manual scavengers
2. To study the process of implementation of the Act, 2013
3. To find out the efforts of the manual scavengers to claim their entitlements

METHODOLOGY

The present study is analytical and empirical in nature and data have been collected from the respondents from the field both in rural and urban areas. The primary data have been collected directly from respondents through focus group discussions and face to face interview. The secondary data have been collected from State and Central Governments records. RTI has been used to collect the data from the government departments. The respondents for the study have been drawn from one gram panchayat, one town panchayat, one municipality, and one corporation. Apart from the above respondents are drawn from one public institution and one Railway station. Apart from the above an intensive interaction with the leader of a civil society organisation who involved in the struggle to liberate the manual scavengers from the inhuman activities for the past one and half decades has been carried out. Basically this study is intended to fine tune the tools and the nuances of data collection for a larger study. This study has a limitation that based on the results and conclusions generalization cannot be made. Yet it will indicate the general trend in the process of implementation of the act. The field data have been collected only from one district, Dindigul and it is one of the backward districts of Tamil Nadu in social development indicators.

DATA ANALYSIS AND INTERPRETATION

RTI Reports

For the details of the manual scavenging, the researcher has filed RTI application in five district of Tamil Nadu, namely Dindigul, Erode, Tirupur, Pudukkottai and Ramanathapuram and got reply. From the reply given to RTI there is no manual scavenger engaged in this profession in all those five districts. The government data show that all the manual scavengers are provided with equipments. Out of the five districts, Pudukkottai district has got only seven manual scavengers but unfortunately the seven manual scavengers are not in the final manual scavengers survey list. Final survey report shows that Pudukkottai district is said to be manual scavenger free district. This kind of confusion persists everywhere in Tamil Nadu.

Government of Tamil Nadu has to constitute a monitoring committee in each district. The District Collector is the head of the committee to monitor the scavengers and the collector has the responsibility to conduct survey properly. But unfortunately the survey of manual scavengers has not been done properly. It has been exposed repeatedly by the media and the civil society organisations involved in rehabilitation of manual scavengers in Tamil Nadu. Because the committee has not made their survey scientifically and transparently so far. As per the government survey report there is no manual scavenger and manual scavenging. If so there is a question who is looking after sewages and drainages? In the meantime, on the 20th January 2016, the Madras High Court gave a direction to the state government that the survey of the manual scavengers was not properly conducted in Tamil Nadu and hence it directed the government to conduct a fresh survey on manual scavengers in response to the petition filed by a social activist Mr. Paadam Narayanan, Director, Change India as a public interest litigation case on the court. He argued that the last survey is over and it informs that only 452 manual scavengers are in the end list in allover Tamil Nadu. But in reality more number of the manual scavengers are in the state. Finally the court issues the order to the state government to conduct fresh survey in the state with all seriousness and with commitment. Now the Government of Tamil Nadu filed fresh affidavits stating that it will conduct the survey only after the conduct of local body elections in October, 2016. The fresh survey will be conducted only after February, 2017. In the meanwhile Change India Director, Narayanan filed an affidavit in the High Court of Madras to pay compensation of Rs. 10 Lakhs to the victims of the family who involved in the sewerage cleaning work as per the Supreme Court direction. The Madras High Court has passed an order the Government of Tamil Nadu to submit a report about the compensation given to the manual scavengers sofar and the steps taken to rehabilitate them. This is the general trend in the process of implementation of the new act in Tamil Nadu.

In this regard, an attempt is made to verify in the field whether the steps indicated and procedures laid down have been followed to enumerate the manual scavengers in Tamil Nadu. An attempt is made to study the conditions of the people who are involved in manual scavenging and for which purposely a few administrative and governance units have been selected. By visiting a gram panchayat, a town panchayat, a municipality, a corporation, a railway station and a public institution, the researcher identified the manual scavengers and interacted with them about their conditions of life and their livelihood. The respondents are drawn from Adipatti village panchayat, Chinnalapatti town panchayat, Odanchattram municipality, Dindigul Municipal Corporation, Dindigul railway station and Gandhigram institutions. Data have been collected from all the above mentioned areas from the Dindigul district. The respondents have been contacted and the research team had interaction with them on their life and livelihood conditions with the help of a check list prepared by the authors of the article. While collecting data, the researchers followed two methods. They are face to face interaction with a few respondents. Focus group discussion had been conducted with all the respondents collectively. Finally the data provided in two settings by the respondents have been verified. While interacting with them it is found out that some of the scavengers belong to backward class. It is strange to notice that a few Muslims have been involved in this work. Never in the media had it surfaced in their reporting. It is only poverty pushed them to this job. The below given tables reveal the nature of work they do how they do and what are all the problems they face in the process. They are being given in the following tables:

Table No. 1 The Caste Wise Distribution of the Respondents

S. No.	Caste	Gender		No. of Respondents	Percentage
		Male	Female		
1.	Scheduled Caste	33	10	43	86
2.	Backward Class	5	2	7	14
Total		**38**	**12**	**50**	**100**

Source: Computed

Table No. 2 The distribution of the respondents based on the type of work they do

S. No.	Type of payment wages	Gender		No. of Respondents	Percentage
		Male	Female		
1.	Contract worker	17	5	22	44
2.	Daily wages	10	6	16	32
3.	Permanent Employee	11	1	12	24
Total		**38**	**12**	**50**	**100**

Source: Computed

Table No. 3 Distribution of the respondents based on the reasons for continuing the jobs

S. No.	Reasons	No. of Respondents	Percentage
1.	Caste Domination	9	18
2.	Official Compulsion	16	32
3.	Don't know Other Work	25	50
	Total	**50**	**100**

Source: Computed

Table No. 4 Distribution of the respondents based on the place of living

S. No.	Place of living	No. of Respondents	Percentage
1.	Village Insight	2	4
2.	Village Outer	15	30
3.	Separate Colony	33	66
	Total	**50**	**100**

Source: Computed

Table No. 5 Distribution of the respondents based on the hereditary occupation or family occupation

S. No.	Hereditary Occupation	No. of Respondents	Percentage
1.	Yes	35	70
2.	No	15	30
	Total	**50**	**100**

Source: Computed

Table No. 6 Distribution of the respondents based on practicing Alcohol

S. No.	Alcohol Practice	Age				No. of Respondents	Percentage
		20-29	30-39	40-49	50-58		
1.	Yes	3	10	16	5	34	68
2.	No	7	5	4	0	16	32
	Total	**10**	**15**	**20**	**5**	**50**	**100**

Source: Computed

Table No. 7 Distribution of the respondent based on awareness about their rehabilitation act

S. No.	Legal Awareness	No. of Respondents	Percentage
1.	Yes	6	12
2.	No	44	88
	Total	**50**	**100**

Source: Computed

Table No. 8 Distribution of the respondents based on usage of modern equipments

S. No.	Usage of Modern Equipments	No. of Respondents	Percentage
1.	Yes	18	18
2.	No	32	64
Total		**50**	**100**

Source: Computed

Table No. 9 Distribution of the respondents based on the discriminatory practice

S. No.	Discrimination	No. of Respondents	Percentage
1.	Tea shops	23	46
2.	Temple	11	22
3.	Other Common places	16	32
Total		**50**	**100**

Source: Computed

Table No. 10 Distribution of the respondents based on attending any training programmes

S. No.	Training	No. of Respondents	Percentage
1.	Yes	12	24
2.	No	38	76
	Total	**50**	**100**

Source: Computed

Table No. 11 Distribution of the respondents based on their name to be enrolled in the manual scavenger's survey

S. No.	Name in the survey	No. of Respondents	Percentage
1.	Yes	-	-
2.	No	50	100
	Total	**50**	**100**

Source: Computed

The above given tables are self explanatory

It is interesting to note that nearly 14% of the respondents belong to backward class and they do scavenging. As they do not have any other skills, they have joined this job and continue this work. They are willing to move out from this work. They do this work not out of love or commitment. They have been drawn to the work only their family socio economic conditions as reported by them. From the above Table No.2 it is inferred that the respondents are performing a number of functions, cleaning the street, cleaning toilets, cleaning drainage and cleaning septic tanks. Scavenging work consists of many tasks and they are to be performed by the same groups. Almost they perform all the works relating to sanitation as multi task workers. Sometimes the work involves serious repercussions from their livelihood perspective and they have not viewed it seriously as their health is at risk. Their perception and understanding of the implications the work they do in their health are very poor. It is a pity in the 70 years of our governance, administration and development activities, nobody has reached them to create awareness of their health and its implications in their life. Only 3 out of 50 respondents involved in septic tank cleaning. The moment they get a call from any household or a local body or an officer, they venture into action. There is no guarantee for their life and yet they do the job. In this process many have given their life. Never had they got compensation. Yet they continue to do that job. In Tamil Nadu there is no instrument or machine for cleaning the septic tanks. Even in a city like Chennai manual workers are involved in this task. In the last year flood in Chennai, the Chennai corporation was not able to complete the cleaning work. They also went to Chennai to clean the city. It is a pity that everybody expressed their concern about the flood victims but nobody expressed their concern about the conditions of the scavengers involved in the work. The respondents made it clear that they are not aware of any instrument or equipment to clean sewerage or septic tanks. They have been performing this work without adequate tools and equipments. They get three types of wages for their work. For contract work they get contract wage. They collect wages daily. They also get monthly salary. But no where they get pension. While asking for reason to stick to this work, the answers are many. They continue this work as they traditionally do the work and from childhood onwards they have been oriented to do the work and they internalized the work. Many do not have any other skill and hence they continue. Of the respondents sixteen replied that the officials called them to work and hence they do this work. Both the workers and the government have no option to move out from this condition. The governments need scientific equipment to do the job. The workers have no other job other than this work. Realistically the issue has to be analysed. Before enacting the legislation, the judiciary could have asked about the preparedness of the state to manage sanitation by using machines. Now all have to escape from the eyes of the law. But at the same time sanitation has to be maintained. 96% of the respondents are far away from the settlement of the rural community. They are either outside the main village or they are living in a separate settlement far away from the main village. It is a pity that the scavengers are living in a sub human living conditions. They clean all the human settlements in the villages and towns and they never clean the areas where they live as they have no time nor they have vision or realization that they live in sub human living conditions. Their settlements are far away from the main villages. 70% of them performed these jobs as hereditary. They have not picked up this work on their own but they have been doing it over generations.

They have been made to internalize the practices what they do for scavenging. Their problems are brought to the forum of the policy community by civil society organisations. But neither the policy community nor the political community evinced concern and interest in their problems as their numerical strength is not much to tilt the election results. Yet the manual scavengers have strong political party affiliation as they nurtured emotional attachment with political leaders. But the community has not been sensitised and as a result this problem continues even today. 68% of them admitted that they take alcohol on daily basis which is necessary for doing this work. Intensity varies from individual to individual. There is no difference between men and women. Quantum also vary from individual to individual. Smoking for men, chewing of aricanut are the two other practices have helped them to do it. They consumed alcohol knowing fully well the implications of alcohol in their health. It is a pity that 88% of the respondents are not aware of the new act and the benefits they get out of implementation of the act. 12% of the respondents are aware of the act that they are clueless to whom they have to approach. 68% of the respondents have some equipment's to perform this duty. It is to be noted here that they are not to protect the people involved in cleaning but to perform the task. They are traditional in nature. Neither the respondents nor the people involve them in these jobs are conscious and sensitive on the importance of the scavengers. All those who are involved in scavenging felt that they are public institutions as the public are aware of the work done by the respondents. They are given the responsibility of cleaning the temples but they are not allowed inside and in the same way they are asked to clean all public place they are not permitted to sit with other in public functions. They are treated as if they are not in the village or in the town to clean it for the benefit of others. Since untouchability is in practice in Tamil Nadu (G. Palanithurai, 2014). It has intensified the discriminatory practices among the people involved in scavenging mostly the Dalits.

Box - I

Hereditary Occupation

Mrs. Virammal is the scavenger of Andipatti village panchayat of Palani Block, Dindigul District. She is 57 years old and belongs to Arunthathiyar community. She is able to read and write. She has owned a house built by the government in the housing schemes of 2001. Now the house is in a dilapidated condition. She has three daughters and all got married. Now she and her husband are living alone. Due to age her husband is not able to work so he is in home. She is not educated and hence she engaged herself with traditional occupation. From early childhood she started learning all the household activities and did her hereditary job. After the death of her father she was forced to do the hereditary job for her livelihood at the age of 18. She started sweeping common places and clean the drainages in her village. She is earning monthly Rs.4000 but compared to agriculture labour her earning is very low.

Non-Dalits are living in the main village and the dalits are living in a separate colony, namely Barathipuram. All non-dalits are having more agricultural lands hence dalits depends on non dalits for their livelihood. They are not allowed to enter into non dalits houses, tea shops, and temples. Village panchayat president post is reserved for dalit women but power is under the control of non dalits. The dalit woman who is in the chair as president of panchayat is performing the role of a manual scavenger. So far Andipatti panchayat is fully controlled by the non dalits community. She has to face discrimination when she enters in the tea shop she use to get tea in a pot and take it at home. Village tea shops having two tumbler system, and dalits who want tea only to take tea, has to stand and take tea. They cannot use the bench provided in the tea stall. Being a panchayat president she has to struggle to sit in the chair in the panchayat office itself.

Even though the village is developed economically toilets users in the village are very minimum in the village. People are using the open space for defecation. The open defecation spot is very near to living places, the panchayat insist her to clean that place. Under dures, she has been asked to clean the open areas. In rainy season very difficult to clean the open defecation place. Usually she cleans the open defecation spots, the human excreta in raw and decomposed manner. Even she cleans drainage with the bare hands without proper equipments. She developed skin allergy and breathing problem. The panchayat does not provide any equipment for drainage cleaning. So the village wastes are dump in the baskets which she carries as head load to disposal location. The waste disposal location is outside the main village. She demands many times for equipments but panchayat officials have not considered her request.

In India there are numerous constitutional provisions to safeguard the dalits and marginalized communities. Particularly the manual scavengers have special rehabilitation Act, the manual scavengers do not know the provisions. The panchayat officials make a record of no manual scavengers in the village panchayat. Because the officials don't know who are the manual scavengers. Till date she is cleaning the village drainages and human excreta from open defecation spot without any safety equipments. But she is working in the name of sanitation worker. She is also the president of the Gram Panchayat.

Box - II

Unable to Move out

Mr. Kannan(37) is the scavenger of Andipattai village panchayat, he belongs to Arunthathiyar community. He has been working as scavenger for last 15 years, with elementary education as his background. He has two children and he hails from a very poor family. He has his own house given by government in the housing schemes. At the age of 22, his father died and the entire family caught in to poverty trap. At that time he went to farm work on daily wages. In the meanwhile one panchayat sweeper passed away, he approached the panchayat president, and president belongs to an upper caste. Kannan easily got that sweeping job at panchayat. His family lives in a separate Aruthathiyar colony. He is earning monthly Rs.4000 but this wage is compared to the wages in farm, it is very low. Since majority belong to caste Hindus, the dalits have been discriminated. One could notice discrimination everywhere from tea stall to temples.

Kannan got married, the family expenditure has gone up. He could not manage the family with the megure amount and hence he wanted to move out from scavenging. But the panchayat president had not accepted his resignation and he continues to do the same work because of the compulsion of the officials. His family lives in separate Aruthathiyar colony. His prime duty is cleaning drainages throughout the villages. He has no equipment's to perform the task.

He came to know about the new act but could not get the due as to how he could avail the benefits he is not aware. As he has some information about the facilities provided by the neighboring Panchayats to the scavenges, he also demanded the same facilities from the panchayat president, but he could not get it as the panchayat revenue could not permit it.

Box - III

No Escape Route

Mr. Alaguraja is a septic tank cleaner of Chinnalapattai town panchayat of Athoor Block, Dindigul District. He is 35 years old and belongs to Kuravar community. His education is only up to 8rd standard. He has four years old son. His parents did some traditional job of Kuravar community. He has no interest in studies and hence he has involved in this job. So his parents decided send his to one septic tank cleaning agency. From early childhood he started learning the septic tank cleaning. He engaged the septic tank cleaning job for his livelihood at the age of 16. He started to clean the septic tanks at Nilakkotai, ten years before he came to Chinnalapattai town and started to clean with his own septic tank vehicle. Under him three people are working in daily wages. He is earning monthly more than Rs.10000.

Every day more than four septic tanks are being cleaned. His septic tank cleaning vehicle was fully modernized one. He has to start the engine and the storage tank would be connected with pipes to lift the solid and water through motor extraction. He is not using any equipments. He has several complications mostly physical. He could not attend to it as he needs mere money to get treatment. Hence he lives on hope with penury. Even though he cleans septic tank with the bare hands without proper equipments. Now she has skin allergy and breathing problem. The septic tank human waste disposal location is outside the panchayat area. No proper location for waste disposal and it is very difficult at that time of rainy season.

The Manual Scavengers have special rehabilitation Act, the manual scavengers do not know the provisions. Those who are working as manual scavengers have to face volley of problems. He doesn't know his position in the society, others are treated as an untouchable. At that time of interaction we told him to

use basic equipment like mask and cloves at that time of cleaning, he has decided himself that manual scavenging is not only profession, it's his life. He doesn't belong to a member of any labour organization. He doesn't know the manual scavengers Act and rehabilitation. But the town panchayat officials have not informed him. The panchayat officials made a record that there is no manual scavengers in the town panchayat. Till dates he is cleaning the septic tank without any safety equipments.

INTEGRATED ANALYSIS

Act has been passed by the Government of India to relieve the manual scavengers from the manual scavenging and to rehabilitate them. But the implementation of the Act lies with the state governments. In the absence of sanitary system supported with science and technology, abolition of manual scavenging is a farse as observed by Mr. Narayanan, Director, Change India, Chennai. Cleaning the villages, towns and cities are in the hands of local bodies assisted by the state governments. The state governments and the local bodies have no assistance from institutions which builds scientific system for scavenging and maintaining sanitation. It is a pity that in the seventy years of the development activities, we have not developed a robust institutional mechanism to evolve technology enabled sanitation and water supply system either by following the western model or evolving an indigenous model (Dhamma Darshan Nigan, 2014). M.K. Gandhi advocated an indigenous model. Still 56% of the population practice open defecation. Tamil Nadu is a fast urbanizing state and yet it has not achieved not even 50% of sanitation through construction of toilets. Even people have not been oriented and sensitized on sanitation and as a result the sanitary conditions are poor both in rural areas and urban cities and towns (Rama Sharma, 1995). Manual scavenging can be abolished through act only when a scientific system is in place for maintaining sanitation. In the absence of the above it is a mere reporting that there is no manual scavenger in Tamil Nadu. But manual scavenging takes place (Government of India, 2016). It is in practice. Civil society organisations have no other go, it has to go to court to exert pressure on the state and local governments. But the state government looks for contribution from the scientific research institutions for creating such a kind of scientific support system for maintaining sanitation (B. Ravichandran, 2011).

From the interaction with the respondents, it is found out that still manual scavenging is going on in rural and urban areas. Conveniently public institutions have handed over the responsibility of maintaining sanitation to the companies. They are not aware of the new act and its implications. Many do not know whether they are paid directly by the institution or through any external agencies. They struggle for their livelihood. The do not have any other skills other this. Even in their areas there is no proper sanitation. They do not have modern equipments to clean the streets and trenches. There is no science and technology based system to manage sanitation both in the rural areas and urban areas. They also face discriminatory practice in the areas where they work. It has been made as a practice to consume alcohol for scavenging work. By doing so, their health conditions are not robust. Railways are no way better.

It is a pity that even at the national level there is no institution to do research on sanitation and support the government to establish a science and technology assisted sanitation system in the country. There is no institutional mechanism to change the mindset and behaviour of the people with regard to sanitation. In the absence of the above we are looking for solution to sanitation and for manual scavengers. It is always a blame game that will not find solution to the problem. We have 765 universities 39000 colleges and 8000000 students are in higher learning institutions. If they are all involved in outreach activities people in 686 districts will be sensitized and conscientised on sanitation. Massively investment has to be made in higher learning institutions to evolve appropriate technology to maintain sanitation scientifically. At present higher learning institutions have not embarked in massive research in sanitation. Thus sanitation and manual scavenging have to be seen integrally to find out solution for open defecation and abolition of manual scavenging.

CONCLUSION

Government of India has enacted a legislation to prohibit manual scavenging and rehabilitate the manual scavengers. This act has come because of the direction given by the Supreme Court. It is to be understood that the implementation requires series of measures not merely allocating money. It requires scientific equipments, technology, management system, people's cooperation. Scavenging is linked with sanitation. Unless sanitation is improved the manual scavengers cannot be rehabilitated.

It is natural that the state governments try to escape from the legal framework as law prohibits the employment of manual scavengers for cleaning try latrine and the places meant for open defecation. But at the same time the government has to discharge the responsibility of maintaining sanitation in the rural and urban areas. In the absence of sanitation behaviour on the part of people and scientific management system to treat the waste scavenging by the scavengers will continue to take place. Further, there are no appropriate tools to be provided to the people who are involved in the job to protect their health and safety of the workers. In this context, one could notice insensitivity on the part of many who involved in management of sanitation. Sanitation work has to be made skill based one and it has to be entrusted to companies. More machinery could be used for cleaning as in the west. Further, scavenging has to be in a more scientific way. Salary for scavenging and cleaning has to be increased. By doing so, it would attract people from different caste groups. Proper rehabilitation work has to be carried out. Now because of the act, people are engaged in such a work but if they met with fatal death, no one takes responsibility. Everywhere one could see manual scavengers, but in record there is no manual scavenger. People have not been oriented for sanitation. There is no sanitation culture among the people. Government fixed targets and construct toilets. But in the absence of sanitation culture, naturally more number of people has to be involved in maintaining sanitation. In the same way, there is no sensitivity on the part of the people about the conditions of the manual scavengers. While interacting with the respondents, it is found out that there is a confusion in identifying the manual scavengers as most of them do

manual scavengers as most of them do multi tasks. Many do not aware of the act and its implications. They are struggling for their livelihood.

In every village panchayat one individual among the elected representatives has to be trained in sanitation and water supply. That individual is brand ambassador for sanitation work in the panchayat areas. Since it is a herculean task achieving clean India through Swatch Bharat scheme, a research institution is need of the hour. It should be a National Institute of Sanitation science and technology.

NOTES AND REFERENCES

1. Chaplin Susan E., (1997). Scavengers: Still Marginalised, **The Administrators**, Vol. XLII, and January – March, 123 - 151.
2. Darshan Nigan Dhamma (2014), A Manual Scavenger, Not your Vote Bank", **Economic and Political Weekly**, Vol. XLIX: No.41 PP 12-13.
3. Government of India, (2016). **Annual report of Department of Social Justice and Empowerment 2015-16,** New Delhi: Ministry of Social Justice and Empowerment.
4. Government of India, (2013).**The Prohibition of Employment as Manual Scavengers and their Rehabilitation Act 2013,** New Delhi: Ministry of Social Justice and Empowerment.
5. Government of India, (1995). **Report of the National Commission for Safai Karmacharis: First Report 1994-95**, New Delhi: Ministry of Social Justice and Empowerment.
6. Government of India, (1996). **Report of the National Commission for Safai Karmacharis: Second Report 1995-96**, New Delhi: Ministry of Social Justice and Empowerment.
7. Government of India, (1998). **Report of the National Commission for Safai Karmacharis: Third Report 1996-97 and 1997 - 98**, New Delhi: Ministry of Social Justice and Empowerment.
8. Government of India, (1999). **Report of the National Commission for Safai Karmacharis: Fourth Report 1999-2000**, New Delhi: Ministry of Social Justice and Empowerment.

9. Kumar Vimal, (2014). Scavenger Community at the Crossroads: Reflection on State Intervention Welfare and Abstruse Welfarism, **Indian Journal of Dalit and Tribal Social Work**, Vol.2:No.1 PP 1-11.

10. Ravichandran B, (2011). Scavenging Profession: Between Class and Caste?" **Economic and Political Weekly**, Vol. XLVI: No.13 PP 21-25.

11. Ramaswami Gita, (2005). **India Striking: Manual Scavengers in Andhra Pradesh and their work**, Pondicherry: Navayana.

13. Siddaramu. B, (2013). Liberation and Rehabilitation of Manual Scavengers, **International Journal of Research in Humanities, Arts and Literature** Vol.1:No.2, July PP29-38.

14. Srivastava B.N., (1997). **Manual scavenging in India: A Disgrace to the Country**, New Delhi: Concept Publishing Company.

15. Singh Bhasha, (2012). **Unseen: The Truth about India's Manual Scavengers** (Translated from Hindi by Reenu Talwar) New Delhi: Penguin Books.

16. Sharma Rama, (1995). **Bangi scaver in Indian Society: Marginality, Identity and Politicization of Community** New Delhi: MD Publications.

17. Samuel Satyaseelan, (2013). Neglect of Sewage Workers: Concerns about the New Act, **Economic and Political Weekly**, Vol. XLVIII: No.49 PP 33-37.

18. Wilson, Bezawada, (2016). Lecture in IIT Madras, 26th.

Chapter IV

Rural Realities and Crises: From the Perception of the Youth Leaders

Youth both educated and the school dropouts in the rural areas are in an ideational dilemma as to what to do at the critical juncture as they face plethora of problems on day today basis at present in Tamil Nadu despite a policy and a planned action[1]. Everyone has aspiration, passion, commitment, skill, capacity and ready to work hard to achieve something[2]. But, they could not find opportunity for meaningful employment to suit their skill, knowledge capacity, and aspirations despite plethora of schemes and programmes for the development of youth[3]. Merit, hard work, commitment, service, skill and dedication have no value and respect in public sphere as corruption favouritism and nepotism have crept in to governance and administration which have eaten away the vital values and probity in public life. Frustration, anger, and dissent loomed large among the youth in Tamil Nadu. It is visible in all protest by the public against the state the presence of the youth as hardcore participants. They involved in all protest movements. They are active in social media. Agitation for Jallikattu is the expression of the all the above. In Tamil Nadu, after the Jalikattu struggle the urban youth started dissenting towards villages and had a pulse of the people who live in villages with minimum for their livelihood.

The conditions of the farmers, farm labourers and others relying on agriculture and animal husbandry for their livelihood made them to feel that the villages are being targeted by the Corporates to make use of the resources[4] and for which the governments are performing a proactive role through their policies and decisions by allowing them to exploit the natural resources recklessly. People in the rural areas on their own started mobilizing and organizing agitations without the support of political parties. People who organize agitations are keeping the political parties away as they have lost their trust on them. In this process, media, political parties, social media spread information both scientific and concocted among the rustic folk. Micro movements are spreading in the rural areas and thereby people get environment awareness. Sensitivity among the people on natural resources is increasing to the level of organising agitations against any kind of exploitation. People have developed restlessness and exhibit anger against the whole system of body politics and administration. Political parties have not been looked at with respect. Politicians are looked at with contempt.

What is going on in rural Tamil Nadu has to be studied quickly. It is not through secondary sources. Hence, an attempt is made to study the rural realities and the prevailing conditions in the rural areas through the perception of rural youth organizers working in different districts all over Tamil Nadu. The prime objective of the study is to assess the perception and attitude of the rural youth as perceived by youth organizers attached with Nehru Yuva Kendra.

METHODOLOGY

A batch of 83 youth volunteers who coordinate the rural youth's activities through Nehru Yuva Kendra participated in a training programme in Gandhigram Rural Institute in the summer June, 2017. The programme was organised for fifteen days to enhance the leadership skills. For this programme resource persons have been drawn from various organisations to interact with the participants. I was also one of the resource persons to deliver lectures. I was asked to brief them about the role of youth in rural local governance and development activities in the rural areas. When I interacted with them, I found many in anger, frustration and restlessness. But they are active, vibrant, serious and sensitive on the problems faced by the people in the rural areas as they live in villages and organise various programmes for rural youth. They asked lot of questions on corruption in politics, governance, administration and development. They further informed that the days are not far off that youth would lose hope in the present system of governance. In such a way, they interacted with me. Having noticed unusual anger among them I wanted to probe further through a focus group discussion. Thus, a focus group discussion was conducted and subsequently an interview schedule was prepared and administered with them. A battery of questions pertinent to politics, governance, administration and development has been asked and they responded all the questions.

RESULTS AND DISCUSSIONS

It is inferred from the focus group discussion that they are totally disgusted over the performance of the ruling class in the state. Many felt that the behaviour of the public representative is natiating and below the level of dignity. Tamil Nadu is a progressive state yet the vitals of politics are eaten away by the political class through their corrupt practices. Tamil Nadu is known for its honest and hard working bureaucracy but now they are co-opted with the corrupt politicians. All have lost the respect from the public. People are the sovereign in a democratic state but in reality they are powerless and most disrespected individuals and they are under the clutches of the government officials. The resources of the villages are not with the people. People in the rural areas are considered as beneficiaries of the government schemes. They are being delivered by the officials. For every service there is a price. Without bribe one cannot get any benefit from the government. What one sees in the village is helplessness and powerlessness which is so serious at present in the rural areas[5]. All the above observations have been generated from the FGD conducted with youth leaders who participated in the training programme in the Gandhigram Rural Institute.

Wide ranges of questions have been asked and for each one they have responded well. For all the questions the respondents have responded and they are being given below:

1. Educational Background

S. No.	Education	Nos	Percentage (%)
1.	School	14	17
2.	College	69	83
	Total	**83**	**100**

2. Age Profile

S. No.	Age	Nos	Percentage (%)
1.	18-21	22	27
2.	22-30	61	73
	Total	**83**	**100**

3. Gender

S. No.	Gender	Nos	Percentage (%)
1.	Male	38	46
2.	Female	45	54
	Total	**83**	**100**

From the above three tables it is inferred that of the respondents most of them are graduates, below the age of 30 and more than half of them are women. Since they are youth organizers, they are well aware of the schemes and programmes of the government both centre and the state for the benefit of the rural poor and more specifically the youth. Periodically they get broachers from centre and state governments about the programmes and schemes implemented in the rural areas for the development of the poor.

4. Do you feel that the enough schemes and programmes are there to solve the problems of rural youth?

S. No.	Answer	Nos	Percentage (%)
1.	Yes	23	28
2.	No	60	72
	Total	**83**	**100**

5. Are you happily leading life by making use of all opportunities in the villages?

S. No.	Answer	Nos	Percentage (%)
1.	Yes	5	6
2	No	78	94
	Total	**83**	**100**

6. Huge youth migration takes place in the villages towards urban areas in search of employment?

S. No.	Answer	Nos	Percentage (%)
1.	Yes	59	71
2.	No	24	29
	Total	**83**	**100**

From the above two tables, it is observed that the present schemes of the government either evolved at the centre or the state are not meeting the requirements of the youth to develop themselves on their own economically. They are mere doles to reduce their distress for a few days[6]. From their perception rural areas have not facilitated the youth to lead a decent human life with dignity as opportunities for development is limited. Youth are always in distress. Youth have to migrate from the rural areas to urban centres for want of employment opportunities. As a result, youth are not contended with what they have in the rural areas as opportunities.

7. Are the rural youth willing to live in the rural areas?

S. No.	Answer	Nos	Percentage (%)
1.	Yes	48	58
2	No	35	42
	Total	**83**	**100**

8. Is there adequate opportunity for development in the villages?

S. No.	Answer	Nos	Percentage (%)
1.	Yes	64	77
2.	No	19	23
	Total	**83**	**100**

Despite the youth's preference to live in the rural areas they are forced to migrate from the rural areas to urban centres for the search of employment opportunities. In the rural areas, there is enough scope to create employment opportunities but there is no institutional and financial mechanism to support any job creation activities. Lot of startup activities could be done by making use of the schemes of Government of India. But, they are not taking place. Skill development and entrepreneurship training have to be conducted for rural youth for gainful employment opportunities. All the institutional facilities created at the dawn of independence for development of agriculture, village industries have been dismantled[7].

9. Are you aware of the process of governance?

S. No.	Answer	Nos	Percentage (%)
1.	Yes	40	48
2.	No	43	52
	Total	**83**	**100**

10. Are you monitoring the Government's performance?

S. No.	Answer	Nos	Percentage (%)
1.	Yes	31	37
2.	No	52	63
	Total	**83**	**100**

Majority feel that it is very difficult to understand the process of governance. It is highly complicated. In the same way, monitoring the performance of the government is also a difficult job. These are all professional exercises and they are not being taught to them[8]. That is why gap exists between people and the government. Awareness level of the people on constitution, governance, administration is very low. They are contended with the beneficiary role. They have to play the stakeholders role and yet they could not do so.

11. Do you feel governance takes place fairly?

S. No.	Answer	Nos	Percentage (%)
1.	Yes	8	10
2.	No	75	90
	Total	**83**	**100**

12. Do you have faith in the governments?

S. No.	Answer	Nos	Percentage (%)
1.	Yes	25	30
2.	No	58	70
	Total	**83**	**100**

13. Do you feel the central government is performing well?

S. No.	Answer	Nos	Percentage (%)
1.	Yes	38	46
2.	No	45	54
	Total	**83**	**100**

14. Do you feel the state government is performing well?

S. No.	Answer	Nos	Percentage (%)
1.	Yes	18	22
2.	No	65	78
	Total	**83**	**100**

They have a feeling that governance performance is not well as they feel fairness is completely lost. Hence, their faith in the government has gone down. When they rate the performance of the state and central government, their rating of the state government is very poor as they closely watch the state government. Yet, they are not feeling better about the performance of the central government. Even academics and researchers could not evaluate the performance of the governments by evolving indicators for measurement. Normally opinion makers help the people to form opinion. Anti-establishment sentiments have been increased at present. It does not mean that they prefer the opposition parties. They hate opposition parties also. They hate the entire political culture and practices.

15. Do you like the present political culture?

S. No.	Answer	Nos	Percentage (%)
1.	Yes	10	12
2.	No	73	88
	Total	**83**	**100**

16. Do you want change in the political culture?

S. No.	Answer	Nos	Percentage (%)
1.	Yes	77	93
2.	No	6	7
	Total	**83**	**100**

17. Do you need change in the central government?

S. No.	Answer	Nos	Percentage (%)
1.	Yes	69	83
2.	No	14	17
	Total	**83**	**100**

18. Do you need change in the state government?

S. No.	Answer	Nos	Percentage (%)
1.	Yes	80	96
2.	No	3	4
	Total	**83**	**100**

They hate the present political culture which is contradictory to democratic culture. Democratic values are slowly eroded[9]. Both the state and the central governments have to be changed. They want a new political force to perpetuate a culture which is pro to democratise the communities. In the rural areas there is no democratic behaviour among the higher caste and class collectivities. They are feudal in their behaviour. Poor and mostly the Dalits longing for equal treatment. Still discriminative process continues in the villages[10]. In the rural areas nobody to listen to the poor. In such a condition poor live in the rural areas[11].

19. Do you feel the rural youth have energy and skill to achieve development?

S. No.	Answer	Nos	Percentage (%)
1.	Yes	78	94
2.	No	5	6
	Total	**83**	**100**

20. Do you feel the education you got will help you to achieve development?

S. No.	Answer	Nos	Percentage (%)
1.	Yes	48	58
2.	No	35	42
	Total	**83**	**100**

21. Do the rural youth have self confidence that they can come up in life?

S. No.	Answer	Nos	Percentage (%)
1.	Yes	75	90
2.	No	8	10
	Total	**83**	**100**

The respondents felt that they have needed aspiration but the education where they have will not create any hope to achieve what they wanted to achieve.

22. Do you feel changes are needed in governance and policies of the government?

S. No.	Answer	Nos	Percentage (%)
1.	Yes	83	100
2.	No	0	0
	Total	**83**	**100**

23. Do you have political affiliation?

S. No.	Answer	Nos	Percentage (%)
1.	Yes	44	53
2.	No	39	47
	Total	**83**	**100**

24. Are you a member in a party?

S. No.	Answer	Nos	Percentage (%)
1.	Yes	22	26
2.	No	61	74
	Total	**83**	**100**

25. Your party

S. No.	Answer	Nos	Percentage (%)
1.	Yes	14	64
2.	No	8	36
	Total	**83**	**100**

All the respondents wanted a total change in policy, governance and politics as they witness the unprecedented distress in the rural areas even though they have political affiliation that too belong to national tradition.

In the focus group discussions, they informed that their responses are not personal but it is a general reflection of the rural youth as they work with rural youth. While working with the rural youth one can infer the powerlessness of the rural people. Day by day people in the rural areas are made vulnerable. They have resources, they have ideas, they have energy, they have hope, they have hard work but they are under the influence and direction of the government. They never feel that they are free. They are always governed by officials. Even after the establishment of the Panchayati Raj system, people never feel that they are free. They feel that the leaders elected by them are under the control of the government officials. They underscored some important points to transform rural India. Rural areas need transformational leaders. They should have passion, commitment, sincerity, sense of dedication and selflessness. The panchayat leaders have to be trained and oriented only in Ashrams as transformational leaders. They should become little Mahatmas. The little Mahatmas can create little republics. For the villages an alternative model of development has to be launched. Their life should be guided by science and they have to follow environment science while they deal with nature. People should be trained to struggle for achieving development. They need social education along with political orientation. They have party orientation not politics but on democracy. We have money, machinery, schemes, programmes, institutions and human resources and they are to be organised properly and they are to be integrated and thereby integral development could be achieved. People should lead a decent, dignified, honourable, scientific, happy and joyful life as citizens. The present perception of beneficiary and petitioner has to disappear. In such a way village life has to be organised.

Box - II

Who leads better life in rural areas?

Lilly (name changed) 23 years old, graduate is working as a passionate coordinator in Nehru Yuva Kendra made some observations about the living conditions of the rural poor in the villages which are to be discussed in appropriate fora. She said that "millions of rupees flowing in the pipeline for development of the poor in the rural areas. Yet, they are not reaching the poor. The poor families did not know how much money flows in the pipeline for development activities. If any one looks at the audited statement of the Gram Panchayat one can easily notice where money goes. Even in MGNREGA alone huge money is spent in each Gram Panchayat. Has that money reached the poor? In the name of poor prosperity has been achieved by both politicians and the officials. If anyone wants to question, his or her life will be made miserable.

All values are only for preaching not for practice". She further lamented that "party men are taught to make money out of individual oriented scheme or group oriented scheme or asset creation in the villages. It is only the politicians and the officials are leading the life with luxury. All others are leading life in subhuman conditions. Life of the poor is always miserable in the villages. People belong to SC will have to face discrimination and they have to lead life without dignity. They have to struggle for getting their legitimate entitlements. Our political parties are the real barriers for the development of the poor. Unless a mass struggle in launched by the villagers, there is no meaningful life for the people who live in the villages".

Box - II

Village Movement

Rajan (name changed) 21 completed his B.A. in social work and participated in youth activities. He informed that "the villages are in the hands of the officials and people have been oriented to act as officials direct to get benefits. Rich households in the rural areas by giving money manipulate everything and get all their entitlements. Officials are in their hands. Officials give respect only to the rich in the villages. Poor have to pay to the officials to get their legitimate entitlements. But poor cannot decide. Panchayati Raj has come but they are being run by a team of the elected representatives and officials as state government wants. People are made to sign in the Gram Sabha documents. Even the Dalit and women representatives could not assert as they have no orientation and understanding of the process of governance. To redeem the villages from the clutches of the dominants and the bureaucracy, a people's movement is the imperative need. He finally said that to reconstruct the villages, people have to be mobilized in Gandhian way. People have to change their attitude and behaviour and by which they can get away from the present consumeristic culture. M.K. Gandhi is more relevant today. Villages could be rejuvenated by adopting Gandhian framework of development even today. Gandhi is needed only now as villages are dump yard to keep the filth of the towns and cities".

CONCLUSION

Prime Minister Narendra Modi wants to create a new India and for which he creates a new consciousness among the people and more particularly among the youth to achieve development. It is called "development consciousness". But in Tamil Nadu the youth are fed up with the way they are governed and administered. Corruption in governance and administration has spoiled the entire public life. Public have been oriented to live in corrupt culture. Valueless politics made people as part of the game. The massive movement for anti-corruption has enabled a party to come to power yet the party has worked to reduce corruption but it regularized the corruption by allowing the parties to collect money legally. As a result, one segment of the people lead life with luxury and another segment has to struggle for getting the minimum requirements to lead a decent dignified human life. Thus, the poor entertain despair instead of hope. They are losing their faith in the political regimes, administration and political process. The finer values, ethics and norms developed by the founding fathers and mothers in India have been made to disappear. This kind of negativity has to be changed. They are looking for a leader who can lead the youth with passion, commitment and honesty to transform Tamil Nadu. They are with pent up feelings and emotions. It may erupt at anytime. When it bust out, it cannot be contained by police. It will have serious repercussions. Series of steps are needed to ease out the frustration of the youth.

NOTES AND REFERENCES

1. Government of India (2014), **Youth Policy 2014**. New Delhi: Ministry of Youth Affairs and Sports.
2. Government of India (2014), **Policy, Schemes and Programmes**. New Delhi: Ministry of Youth Affairs and Sports.
3. Government of India (2017), **Youth in India-2017**. New Delhi: Ministry of Statistics and Programme Implementation.
4. M. V. Nadkarni (2018). Crises in Agriculture: Can it be overcome? **Economic and Political Weekly**, Vol. LIII: No.17, April 28.
5. It is not only in Tamil Nadu but also in other states. Hari K. Nagarajan, Hans P. Binswanger- Mkhize and S.S. Meenakshisundaram (2015), **Decentralization and Empowerment for Rural Development**. New Delhi: Foundation Books.

6. Giving something to the poor is not a development initiative. Robert Chamber (1983). **Rural Development: Putting the Last First**. Harlow: Longman Scientific and Technical.

7. G. Palanithurai and R.Ramesh (2008), **Globalization: Issues at the Grassroots**. New Delhi: Concept Publishing Company.

8. Karunakar Patnaik (2017), **Governance Reforms: Critical Issues and Challenges**. New Delhi: Concept Publishing Company.

9. G. Palanithurai (1995), **People's Perception of Politics: An Indian Perspective**. New Delhi: Konark Publishers and Distributors.

10. G. Palanithurai, "Status of Dalits and Dalit Gram Panchayat Leaders in Tamil Nadu" in Rakesh Hooja and Sunil Dutt (eds) (2010), **Fifty Years of Panchayati Raj and Decentralized Development.** New Delhi: Indian Institute of Public Administration.

11. S. Parasuraman, Gomathy, Kumaran Raj and Bina Fernandez (2003), **Listening to People Living in Poverty**. Bangalore: Books for Change.

Conclusion

It is to be noted that India has a detailed constitution and more number of pro-poor laws in the world. Entitlements of the poor and marginalised are made as Acts and huge outlays for implementation of the acts. But in reality, most of them are in document. The allocated amount for implementation of schemes and programs meant for poor has not reached the targeted group. In India, we have democracy but no democrats. After Independence, the expected social revolution envisaged in the constitution has not been achieved. Only very rarely we have seen such revolutions. Kerala is one which has witnessed such a revolution. Our political class has lost its morality as they are sinking in corruption and vote bank politics. Our bureaucracy has worked for itself in the name of the people. In India power of the system works very well but power of the people could not work as people are disempowered through the state continuously. Small farmers, farm laborers, small traders, street vendors, nomads, fisher folk, tribals, crafts men and scavengers have been facing plethora of problems. They constitute more than 70% of the population and yet they could not find solutions to the problems. The existing gap between the government and the poor is so vast. Poor believe that they are under the control of the state. They believe that they cannot question. Government believes that people are the beneficiaries and government has to dole out benefits. The process of governance is much complex and complicated and it cannot be understood by the poor. This information gap is advantageous for the ruling elite and bureaucracy. Indian society appears to be in democracy. But in reality it is institutional in nature not behavioural. Hence, democracy works in hostile conditions. As a result, what could be achieved through democracy has not been achieved?

Whether to reduce poverty or to redeem from discrimination of the poor have to be organised, educated, informed, sensitized and empowered to claim their entitlements as their rights. While claiming their entitlements, the poor should not feel that they are at the receiving end as beneficiaries. They first establish their equality and claim their entitlements. Democratization of the community is the first step. Only by creating a sense of equality, poor can claim their entitlements. It is a process of struggle. The poor and the marginalised have to be trained in using the Acts, and schemes for their development. At the same time, the process of governance has to be simplified to the level of the understanding of the poor. Poor has to be trained in handling of the democratic methods in the governance institutions to claim their entitlements. It is to be understood that all the pro-poor Acts and Schemes have come into reality through a relentless struggle and that struggle has to continue to make use of the historic Acts and schemes. Thus empowerment of the poor, democratization of communities and simplifying the process of governance will go a long way to establish a just equalitarian society.

Bibliography

1. Ahmed Sadig (2007), **Job Creation and Poverty Reduction in India: Towards Rapid Sustained Economic Growth**. New Delhi: Sage Publications.

2. Austin Grenville (2012), **Working a Democratic Constitution: A History of the Indian Experience** New Delhi: Oxford University Press.

3. Austin Grenville (1994), **The Indian Constitutions' Seamless Web**. RGICS Paper No.15. New Delhi: Rajiv Gandhi Foundation.

4. Basu Amrita and Kholi Atul (eds) (1998), Community Conflicts and the State in India. New Delhi: Oxford University Press.

5. Bhatt, Ela (1989), **Shramshakti: Report of the National Commission on Self-Employed Woman and Women in the Informal Sector**. New Delhi: Government of India.

6. Chamber Robert (1983), **Rural Development: Putting the Last First**. Harlow: Longman Scientific and Technical.

7. Chaplin Susan E., (1997), Scavengers: Still Marginalised, **The Administrators**, Vol. XLII, and January – March.

8. Charmaz Kathy (1983), "The Grounded Theory Method: An Explication and Interpretation" in **Contemporary Field Research** (eds) Emerson R. M. Boston: Little Brown.

9. Coelho Karen, Venkat T and Chandira K (2013), "Housing, Homes and Domestic Work: A Study of Paid Domestic workers from a Resettlement Colony in Chennai" **Economic and Political Weekly**, Vol. XLVIII: NO. 43.

32. Nagarajan Hari K., Binswanger- Mkhize Hans P. and Meenakshisundaram S.S. (2015), **Decentralization and Empowerment for Rural Development**. New Delhi: Foundation Books.

33. Neetha, N (2013) "Minimum Wage of Domestic Workers: Mirroring Devalued House Work, **Economic and Political Weekly**, 48 :(43).

34. Neetha. N (2008), "Regulating Domestic Work", **Economic and Political Weekly**, 43(37).

35. Palanithurai G. (1995), **People's Perception of Politics: An Indian Perspective**. New Delhi: Konark Publishers and Distributors.

36. Palanithurai G. and Ramesh R. (2008), **Globalization: Issues at the Grassroots**. New Delhi: Concept Publishing Company.

37. Palanithurai G., "Status of Dalits and Dalit Gram Panchayat Leaders in Tamil Nadu" in Hooja Rakesh and Dutt Sunil (eds) (2010), **Fifty Years of Panchayati Raj and Decentralized Development.** New Delhi: Indian Institute of Public Administration.

38. Parasuraman S., Gomathy, Raj Kumaran and Fernandez Bina (2003), **Listening to People Living in Poverty**. Bangalore: Books for Change.

39. Patnaik Karunakar (2017), **Governance Reforms: Critical Issues and Challenges**. New Delhi: Concept Publishing Company.

40. Ramaswami Gita, (2005), **India Striking: Manual Scavengers in Andhra Pradesh and their work**, Pondicherry: Navayana.

41. Ravichandran B, (2011), Scavenging Profession: Between Class and Caste?" **Economic and Political Weekly**, Vol. XLVI: No.13.

42. Samuel Satyaseelan, (2013), Neglect of Sewage Workers: Concerns about the New Act, **Economic and Political Weekly**, Vol. XLVIII: No.49.

43. Sankaran, Kamala (2013), "Domestic Work, Unpaid Work and Wage Rates" **Economic and Political Weekly**, Vol. XLVIII: No. 43.

44. Senthil Natarajan D. (2016), "Two women oversee burial ground work" **Deccan Chronicle** 25th August.

45. Sharma Rama, (1995), **Bangi scaver in Indian Society: Marginality, Identity and Politicization of Community** New Delhi: MD Publications.

46. Siddaramu. B, (2013), Liberation and Rehabilitation of Manual Scavengers, **International Journal of Research in Humanities, Arts and Literature** Vol.1:No.2, July.

47. Singh Bhasha (2014), **Unseen: The truth about Indian's Manual Scavengers**. New Delhi: Penguin Books.

48. Srivastava B.N. (1997), **Manual Scavenging in India: A Disgrace to the Country**. New Delhi: Concept Publishing Company.

49. Staff Reporter (2013), "Burning the body in the night by women vettiyan" (in Tamil) **Dinamalar** Tamil Daily, 9th June.

50. Staff Reporter (2017),"A Women in Challenging task at Graveyard" **Social** 8th March.

51. Tamil Nadu Mayana Vettiyan Sangam (2009), **Epic Age to the Present: The Life of the Mayana Vettiyans A Report** Madurai: Tamil Nadu Labour Union.

52. Vadageri, Bhagyashree. S and Soundari M. Hilaria (2016), "Socio-Economic Problems of Domestic Labourers in Karnataka" **Indian Journal of Economics and Development** 4:(4).

53. Vannan Gokul (2014), "Grave Digger Takes Green Route to Mark Amavasya at Madurai Cremation Ground" **The New Indian Express** 24th September.

54. Venkat Vidya (2008), "They deal with death every day at work" **The Hindu** 24th August,.

Index

W

Y

Made in United States
Troutdale, OR
11/03/2026

56844628R10069